雁荡山联合国教科文组织世界地质公园
科学导游手册

SCIENTIFIC TOURISM GUIDEBOOK OF YANDANGSHAN UNESCO GLOBAL GEOPARK

袁茂珂 杨 林 卢琴飞 等 编著

中国地质大学出版社
CHINA UNIVERSITY OF GEOSCIENCES PRESS

图书在版编目(CIP)数据

雁荡山联合国教科文组织世界地质公园科学导游手册/袁茂珂等编著. —武汉:中国地质大学出版社,2021.8

ISBN 978–7–5625–4973–4

Ⅰ.①雁…

Ⅱ.①袁…

Ⅲ.①雁荡山–地质–国家公园–旅游指南

Ⅳ.①S759.93

中国版本图书馆CIP数据核字(2021)第149382号

雁荡山联合国教科文组织世界地质公园
科学导游手册

袁茂珂 杨 林 卢琴飞 等 编著

责任编辑:胡珞兰 龙昭月	选题策划:胡珞兰 毕克成 段 勇	责任校对:何澍语
出版发行:中国地质大学出版社(武汉市洪山区鲁磨路388号)		邮政编码:430074
电 话:(027)67883511 传 真:(027)67883580		E–mail:cbb@cug.edu.cn
经 销:全国新华书店		http://cugp.cug.edu.cn
开本:880毫米×1230毫米 1/24		字数:123千字 印张:4.25
版次:2021年8月第1版		印次:2021年8月第1次印刷
印刷:湖北新华印务有限公司		印数:1 — 5300册
ISBN 978–7–5625–4973–4		定价:38.00元

如有印装质量问题请与印刷厂联系调换

《雁荡山联合国教科文组织世界地质公园科学导游手册》编委会
SCIENTIFIC TOURISM GUIDEBOOK OF YANDANGSHAN UNESCO GLOBAL GEOPARK Editorial Committee

总策划 Chief Planner	黄　靖 Huang Jing
策　划 Planner	黄升良　狄永明　胡约素 Huang Shengliang　Di Yongming　Hu Yuesu
编委会委员 Chief Editor	袁茂珂　杨　林　卢琴飞　张成功　宁萌萌 Yuan Maoke　Yang Lin　Lu Qinfei　Zhang Chenggong　Ning Mengmeng
译　者 Translator	吴振扬 Wu Zhenyang
摄　影 Photographer	叶金涛　等 Ye Jintao etc.
设　计 Designer	王丽君 Wang Lijun

目 录
Contents

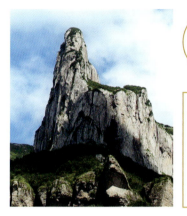

雁荡山联合国教科文组织世界地质公园简介
Introduction to Yandangshan UNESCO Global Geopark / 01

自然地理　Physical Geography　　　　　　　　　　／ 05
　地理位置 Geographical Position　　　　　　　　　／ 06
　地形地貌 Landform　　　　　　　　　　　　　　 ／ 06
　气候条件 Climatic Conditions　　　　　　　　　　／ 07
　公园特色 Characteristics of the Geopark　　　　　　／ 08

地质遗迹景观 Geological Relic Landscape　　　　　　／ 13
　雁荡山园区 Yandangshan Scenic District　　　　　　／ 14
　方山－长屿硐天园区 Fangshan-Changyu Dongtian Scenic District / 39
　楠溪江园区 Nanxijiang Scenic District　　　　　　　／ 48

人文景点推荐 Recommended Cultural Scenic Spots　　／ 53
　章纶墓 Zhanglun Tomb　　　　　　　　　　　　　／ 54
　南阁牌坊群 Nange Memorial Archway Group　　　　／ 56
　灵岩禅寺 Lingyan Temple　　　　　　　　　　　　／ 57
　雁荡山摩崖石刻 Yandangshan Inscriptions on Precipices ／ 58
　岩头古村 Yantou Ancient Village　　　　　　　　　／ 59
　苍坡古村 Cangpo Ancient Village　　　　　　　　　／ 61

雁荡山世界地质公园科学故事
Science Stories of Yandangshan UNESCO Global Geopark / 63

什么是火山？ What are Volcanoes? / 64
惊天动地的火山爆发景观 Earth-Shaking Scenery of Volcanic Explosion / 65
火山岩浆喷溢地表的景观 Effusion Spectacle of Volcanic Magma / 67
岩石地层柱与同位素年龄——雁荡山火山真实年龄
Lithostratigraphic Column and Isotopic Age—Age of Yandangshan Caldera / 69
火山家族 Volcano Family / 71
什么是破火山？ What is a Caldera? / 73
什么是火山岩？ What are Volcanic Rocks? / 74
雁荡山：火山岩的天然博物馆 Yandangshan: A Natural Museum of Volcanic Rocks / 75
雁荡山火山生命史——雁荡山火山演化模型 Evolution History of Yandangshan Caldera—Evolution Model of Yandangshan Caldera / 76
雁荡山火山由来 The Origin of Yandangshan Caldera / 78
雁荡山火山喷发后的地质作用——大自然的力量塑造了雁荡山"不类他山"的风光
The Post Volcanic Eruption Geological Processes—The Power of Nature Creating the Uniqueness of Yandangshan Scenery / 79
雁荡山嶂、门、柱的形成 Formation of Yandangshan Cliffs, Gates, and Columns / 81

旅游服务 Tourist Services / 83

雁荡山世界地质公园博物馆
The Museum of Yandangshan UNESCO Global Geopark / 84
旅游资讯 Tourist Information / 85
  吃 Food / 85
  住 Hotel / 88
  行 Transportation / 89
  游 Travel / 91
  购 Shopping / 92
常用电话 Frequently Used Telephone Numbers / 93
旅游小贴士 Travelling Tips / 93

雁荡山联合国教科文组织世界地质公园简介

雁荡山联合国教科文组织世界地质公园(以下简称为雁荡山世界地质公园)位于浙江省温州市与台州市,由雁荡山、楠溪江、方山-长屿硐天3个各具特色的园区组成,面积298.80平方千米,是一个以火山岩地质地貌为主导,千年宗教历史、山水文化和石文化交相辉映的综合性自然公园。1982年被国务院列为首批国家重点风景名胜区,2004年被评为国家地质公园,2005年被评为世界地质公园,2007年被评列为国家首批AAAAA级旅游景区,2010年被评为国家矿山公园。

雁荡山是中国著名的"三山五岳"之一,在地理位置上,地处西太平洋亚洲大陆边缘,是全球巨型火山带上最具代表性的古火山之一,发育了具有全球意义的、典型的、类型丰富的火山岩和火山岩地貌。从大约1.4亿年前开始,古太平洋板块向亚洲大陆板块俯冲,在亚洲大陆板块边缘形成了大规模的火山和岩浆侵入活动。随后雁荡山经历了4期火山爆发,后经地质改造,形成了如今的奇峰叠嶂、怪石飞瀑。这些自然景观的形成过程完整记录了古火山喷发、演化历史,为人们揭示了1亿多年前火山爆发的奥秘。

Introduction to Yandangshan UNESCO Global Geopark

Yandangshan UNESCO Global Geopark is located in the cities of Wenzhou and Taizhou in Zhejiang Province. It comprises Yandangshan, Nanxijiang, Fangshan-Changyu Dongtian (Changyu Cave) with a total area of 298.80km². It is dominated by volcanic geology and landscape, integrated with religious history, cultural landscape and stone culture to form a comprehensive nature park. The geopark was Listed in the first batch of National Park of China in 1982. It was approved as National Geopark in 2004, a member of the Global Geoparks Network in 2005 and one of the first AAAAA National Scenic Attractions in China in 2007 and obtained the title of "National Mining Park" in 2010.

Yandangshan is one of the most famous "three hills and five moutains" of China. It is located at the western end of the Asia-Pacific continental margin. It is a gigantic ancient volcano with global representation and significance in terms of the different types of volcanic rocks and landforms. About 140 million years ago, the Paleo-Pacific plate subducted beneath the continental plate to create a large-scale volcanic and magmatic intrusion on the edge of the continental plate. Yandangshan has experienced four stages of volcano eruption and undergone tremendous geological transformation, resulted in its towering peaks, peculiar rocks and waterfalls. The geology and natural landscape have recorded the evolution of the ancient volcano, revealing the mystery of its eruption and changes more than 100 million years ago.

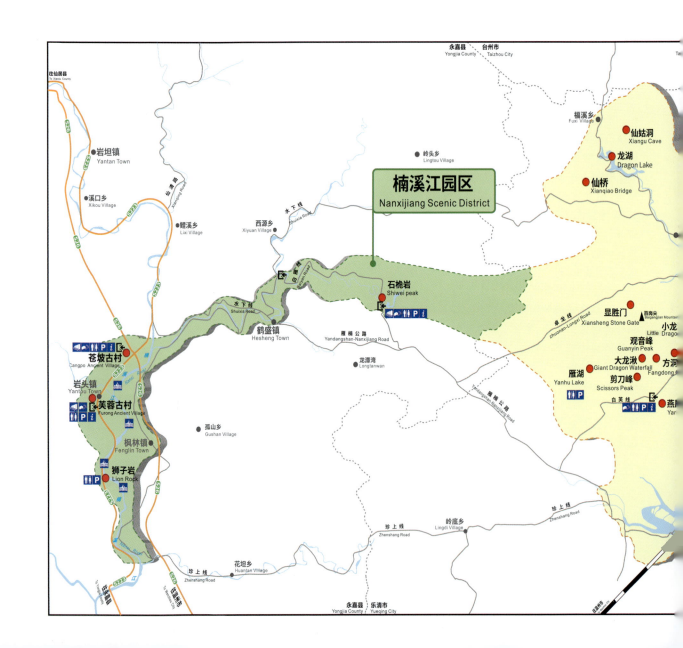

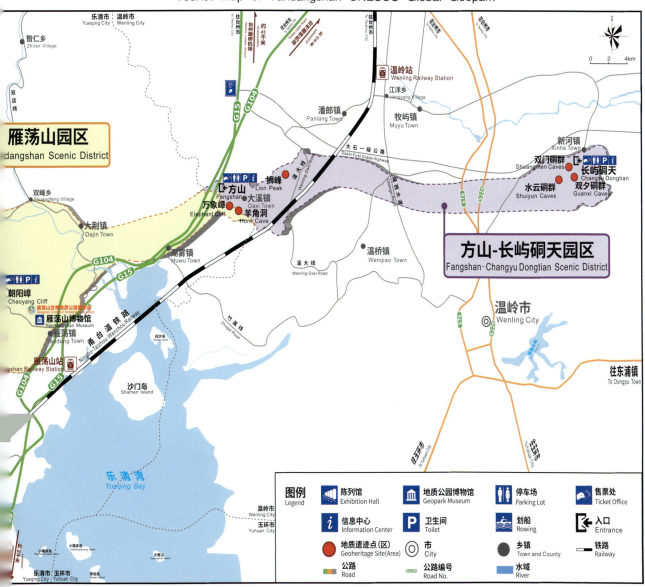

自然地理

Physical Geography

地理位置

雁荡山世界地质公园位于中国浙江省温州市和台州市境内，东临乐清湾，西达永嘉县，北接台州市，南濒乐清市。地理坐标：东经120°41′40″—121°27′40″，北纬28°12′30″—28°30′00″。

Geographical Position

Yandangshan UNESCO Global Geopark is located in Wenzhou and Taizhou cities in Zhejiang Province which faces Yueqing Bay in the east, extends to Yongjia County in the west, joins Taizhou City in the north and is bounded by Yueqing City in the south. Its geographic coordinates are E120°41′40″－121°27′40″ and N28°12′30″－28°30′00″。

Landform

Yandangshan UNESCO Global Geopark, part of the southern branches of Kuocang Mountain, is composed of Yanhujian, Baigangjian, Lingyunjian, Wuyanjian and other peaks or hills. With Yueqing Bay in the east, Yandangshan is crowned as "the famous mountain on the sea". Extending from the northeast to southwest, the landforms are characterized by outstanding altitude differences. It is 600－800m above sea level and can be classified as low-relief terrain. However, its highest peak is up to 1108m above sea level.

地形地貌

雁荡山世界地质公园位于括苍山南缘支脉，由雁湖尖、百岗尖、凌云尖、乌岩尖等峰峦组成，东临乐清湾，有"海上名山"之美誉。山势西高东低，呈北东-南西走向，高低悬殊，海拔600~800米，属低山区，最高峰海拔1108米。

气候条件

雁荡山世界地质公园地处东海之滨，毗邻乐清湾，属亚热带海洋性季风气候，雨量充沛、气候温暖，冬无严寒，夏无酷暑。低温期短，无霜期长，平均无霜期269天。年平均气温13.5℃，最冷1月份平均气温5~7℃，最热7月份平均气温27℃。平均年降雨量1936毫米，年降雨量最高达2127毫米，年平均相对湿度为77%。

公园气候宜人，树木葱郁，空气清新，是科学考察、休闲度假、观光旅游之胜地。

Climatic Conditions

Yandangshan UNESCO Global Geopark borders the East China Sea and Yueqing Bay. It is situated in the subtropical oceanic climate zone which has abundant rainfalls, warm temperature, mild winter and free from intense summer heat. Besides, the cold period is relatively short with long frost-free winter. Its annual average temperature is 13.5℃. In January and July, the coldest month and the hottest month in a year, the average temperature is 5-7℃ and 27℃ respectively and has a total of 269 frost-free days. The average annual rainfall is 1936mm with a maximum value of up to 2127mm and an average relative humidity of 77%.

With the pleasant temperature, luxuriantly green vegetation and crystal clear air, the geopark is the best site for scientific research as well as for sightseeing and leisure on holidays.

公园特色

Characteristics of the Geopark

雁荡山世界地质公园之美在于以奇秀为本，险峻、幽奥、旷远并蓄。以叠嶂锐峰、奇岩怪洞、门阙岩岗为骨骼，以飞瀑、涧溪、湖潭为脉搏，两者配置和谐，结构独特，气势磅礴，不愧为"天下奇秀"。

Yandangshan UNESCO Global Geopark is elegantly beautiful without the sacrifice of precipitous, peaceful and long-standing sceneries. The peaks raise one higher than another and are inundated with remarkable rock features and caves. Besides, waters of waterfalls, streams and lakes flow gracefully, etherealizing the entire region by highlighting its unique topography.

海上名山——雁荡山 The Famous Mountain on the Sea—Yandangshan

自然地理
Physical Geography

雁荡山是中生代晚期破火山的典型代表,是一部亚洲大陆边缘白垩纪时期破火山形成与演化的永久性文献,也是研究酸性岩浆作用的流纹岩天然博物馆。这里遍布古太平洋板块与亚洲大陆板块相互作用动力学过程的火山学与岩石学证据,发育了叠嶂、奇峰、怪洞等典型滨海流纹岩地貌。

Yandangshan is a typical caldera in the late stage of Mesozoic and a permanent record of the formation and evolution history of caldera at the edge of Asian continent in Cretaceous period. It is a natural museum of rhyolites formed due to acidic lava. Geological evidences are distributed extensively at every corner of the geopark, offering the convenience of studying volcanoes and its associated processes which took place at the margin of the Paleo-Pacific plate and the Asian continental plate. The geopark is characterized by rising hills, steep peaks, attractive caves and other typical coastal rhyolitic landforms.

雁荡山不仅发育流纹质火山岩地貌,还发育独特的角砾凝灰岩采石遗址。长屿硐天有长达1500多年的采石历史,形成了28个硐群,1314个形态各异的硐窟,可谓是"虽由人作,宛自天开"。

Apart from the rhyolitic landforms, Yandangshan also has historic quarry sites of brecciated tuff. Some sites have a history of over 1500 years. The Changyu-Dongtian site has 28 cave groups made up of 1314 individual smaller caves of different shapes. They are regarded as "naturally created man-made caves".

金带嶂 Golden Belt Cliff

流纹岩 Rhyolite

硐天花园 Dongtian Garden

观夕硐 Guanxi Cave

雁荡山是文人的天堂,也是文人的故乡,完整的楠溪江水系、历史悠久的古村落群,共同阐释了这里"水美、岩奇、瀑多、林秀、村古"的内涵,是中国农耕文明与山水文化完美结合的杰作。永嘉学派、永嘉四灵、永嘉昆曲,书写了中国文化史上光辉灿烂的篇章。

Yandangshan is a paradise and hometown of scholars. The integration of Nanxijiang system, the historical villages, the natural water features, peculiar rocks, waterfalls and forests has prepared the area for a perfect breeding ground for the traditional Chinese farm-studying and landscape cultures. They are well-represented by the proliferation of Yongjia School of Thought, Yongjia Four Poets and Yongjia Kun Opera which have played a significant role in the development of Chinese cultural history.

楠溪江　Nanxijiang

苍坡古村　Cangpo Ancient Village

白鹇　*Lophura nycthemera*　　　猕猴　*Macaca mulatta*

自然地理
Physical Geography

雁荡山优美的自然环境孕育了丰富的动植物资源。它们的存在不仅改造并装点着雁荡山，还为雁荡山带来了生机与活力。

The beautiful natural environment of Yandangshan has enhanced the biodiversity of this area. The abundant fauna and flora not only enrich but also invigorate and vitalize Yandangshan.

杰出的科学价值、奇秀的自然美景、浓厚的山水文化、悠久的人文历史，共同组成了大自然的奇观、世界的宝贵遗产。

Yandangshan's great scientific value, natural beauty, rich landscape culture and long history have integrated to constitute the wonders of nature and the precious heritage of the world.

鸳鸯　*Aix galericulata*

华西枫杨　*Pterocarya macroptera* var. *insignis*

鹅掌楸　*Liriodendron chinense*（Hemsl.）Sarg.

南方红豆杉　*Taxus wallichiana* var. *mairei*

地质遗迹景观

Geological Relic Landscape

雁荡山园区

    雁荡山园区以山奇水秀闻名，素有"海上名山、寰中绝胜"之誉，史称中国"东南第一山"。它以白垩纪（1.45亿～0.66亿年）的流纹质火山岩地貌为基础而形成奇峰怪石、飞瀑叠嶂。园区主要包括灵峰、灵岩、大龙湫、三折瀑、雁湖、显胜门、仙桥、羊角洞八大景区，其中灵峰、灵岩、大龙湫景区由于其优美的自然风光被誉为"二灵一龙"。这里也曾是第三十届国际地质大会的考察地之一，来自全球多个国家的地质学家对雁荡山教科书般的流纹质火山岩地貌惊叹不已。

Yandangshan Scenic District

    Yandangshan Scenic District is famous for its mountainous landscape and is well-known as "the No.1 mountain in Southeast China". The scenic district is dominated by Cretaceous (a geological age of 145 – 66 million years) rhyolitic volcanic rocks and landform. It comprises eight different scenic geosites including Lingfeng, Lingyan, Dalongqiu, Sanzhepu, Yanhu, Xianshengmen, Xianqiao and Yangjiaodong. Lingfeng, Lingyan and Dalongqiu in particular, which have been named "Two ling (souls) and one dragon", are very well-known to visitors. The scenic district was inspected by geologicial scholars worldwide attending the 30th International Geological Congress and its distinctive rhyolitic volcanic geomorphology had greatly impressed them.

灵峰晨雾
Morning Fog of Lingfeng

海上名山　Famous Mountain on the Sea

朝阳嶂

朝阳嶂位于灵峰景区入口,横亘400余米,高百米以上,层层叠叠,方展如屏。明代著名地理学家、旅行家徐霞客(1587—1641)将这种山体称之为"叠嶂"。朝阳嶂的岩石为雁荡山火山第三期多次喷溢出的熔岩,冷却后形成的流纹岩经断裂切割形成岩嶂。山体下部为含角砾、石泡的流纹岩,易剥落成为洞穴。

Chaoyang Cliff

Seated at the entrance of Lingfeng Scenic Area, Chaoyang Cliff is about over 400m wide and over 100m high. It stretches upward layers by layers and looks like a hanging curtain. Xu Xiake (1587 – 1641), the famous geographer and traveller of Ming Dynasty, described it as "cliffs". Rocks of Chaoyang Cliff were formed during the 3rd stage of volcanic eruption at Yandangshan when lava gradually cooled down before turning into rhyolite. Later, the rocks were cut by faulting to create cliff. The lower section of the cliff is made up of breccia (angular fragments) and lithophysa (globular) rhyolite and can easily desquamate to form holes and caves due to weathering.

朝阳嶂　Chaoyang Cliff

灵峰古洞

灵峰古洞位于灵峰景区中部，鸣玉溪畔，俗称"倒灵峰"。天圣元年（公元1023年）曾在此建寺，距今600多年前的元代，发生了一次地质灾害——因地质作用引发山体岩石崩落，崩落的巨石在山角处架空堆积成洞穴群。令人惊奇的是，巨石架空形成的洞穴形态各异，洞洞相连，迂回曲折，深幽奇特，构成了云雾、透天、含珠、隐虎、好运、玲珑、凉风等七洞。

Lingfeng Ancient Cave

In the middle of Lingfeng Scenic Area, Lingfeng Ancient Cave is located by the side of Mingyu Stream and is usually being known as "Inverted Lingfeng". Temples were built here since the first year of the Tiansheng Era（A.D.1023）, but were destroyed by landslide and rock fall about 600 years ago during the Yuan Dynasty. As a result, gigantic boulders were piled up to form caves. These caves were surprisingly connected with each other in different forms and shapes to create seven famous caves called Yunwu, Toutian, Hanzhu, Yinhu, Haoyun, Linglong, and Liangfeng.

灵峰古洞 Lingfeng Ancient Cave

合掌峰

合掌峰是灵峰景区的核心景观，位于鸣玉溪畔。合掌峰是雁荡山代表景观之一，峰高约270米，在群峰环拱中直插云霄。该峰岩石为雁荡山火山第二期喷溢的流纹岩，后因断裂作用，一峰开裂为二，左称灵峰，右称倚天峰，状若合十的手掌，两掌之间即为观音洞，洞高113米，深76米，宽14米。九层楼阁倚洞就势而建。建筑与洞穴完美结合，巧夺天工，国内外建筑人士倍加赞赏。邓拓（1912—1966）诗曰"两峰合掌即仙乡，九叠危楼洞里藏"。洞内有"洗心""漱玉""石釜"三泉，洞内阁楼是观灵峰全景的最佳处之一。

Hezhang (Clasped Palm) Peak

Hezhang (Clasped Palm) Peak is the most important scenic spot in Lingfeng Scenic Area, located near the Mingyu Stream. It is an iconic symbol of the Yandangshan Scenic District. The peak, surrounded by hills, is 270m high and rises up into the clouds. It was formed during the 2nd stage of volcanic eruption at Yandangshan and later separated by faulting. When being viewed together, Lingfeng Peak and Yitian Peak look like two clasped palms and between them is the Guanyin Cave. The Guanyin Cave is 113m high, 76m deep and 14m wide. To make use of available space of the cave, a nine-storey Buddhist temple was built to fit well with the surrounding environment. The architecture has been highly commended by architects at home and abroad. Deng Tuo, for example, wrote in his poem that "Two peaks being together to form the Hezhang Peak (a Buddhist fairyland) with the hidden nine-storey Buddhist temple inside the Guanyin Cave". There are also three springs inside the cave, namely Xixin, Shuyu, and Shifu. The attic in the cave is the best spot to have a bird's-eye view of the entire Lingfeng Scenic Area.

观音洞　Guanyin Cave

合掌峰　Hezhang (Clasped Palm) Peak

灵峰夜景

灵峰夜景堪称"雁荡一绝"。每当夜幕降临时，白天看似普普通通的山峰都会披上神秘的盛装，惟妙惟肖，如入仙境一般。灵峰"日景耐看，夜景销魂"，因为这些奇峰怪石在月光和夜色的映衬下，犹如涂上了神秘而温馨的色彩，构成了一幅幅线条鲜明的泼墨画，勾画出一张张美丽的剪影，使灵峰夜景更具有形象美、意境美。

Night Scene in Lingfeng

Night scene in Lingfeng is one of the best attractions in Yandangshan. The peaks look common during daytime but charming when night comes. When the sky turns dark and the peaks are bathed in moonlight, the landscapes are painted with mysterious but warm colours to produce graceful splash ink paintings and silhouettes. All of these have added beauty to the image and artistic conception to the attraction.

雄鹰展翅（夜景） Eagle Spreading Wings (Night Scene)

犀牛峰（夜景） Rhino Peak (Night Scene)

双乳峰（夜景） Shuangru Peak (Night Scene)

合掌峰（夜景）
Hezhang (Clasped Palm) Peak (Night Scene)

中折瀑

中折瀑位于三折瀑景区。瀑布落差120余米，飞瀑从半圆形洞顶凌空而下，状如帘珠，阳光照之，彩虹飞动，令人叫绝。郭沫若（1892—1978年）诗赞："我爱中折瀑，珠帘掩翠楼。"瀑壁呈半圆桶状，因而有人说中折瀑为火山口，其实它不是火山口，而是火山喷溢流纹岩的纹理形成的。

Zhongzhe Waterfall

Zhongzhe Waterfall is located in Sanzhepu Scenic Area. The water falls down like a bead curtain from a semicircular cave top of over 120m high. Under sunshine, the rainbows appear in the sky and add attraction to the waterfall. Guo Moruo (1892 – 1978) once complimented in his poem that "I love Zhongzhe Waterfall as the bead curtain hides the green tower". Some people think that its semicircular shape is a volcanic crater, but in fact it is not. It is caused by the texture of rhyolite.

中折瀑　Zhongzhe Waterfall

铁城嶂

铁城嶂位于三折瀑景区净名谷中。铁城嶂"势若长城,色如黑铁",与游丝嶂对峙,巍然奇诡,被称为"雁荡山第一叠嶂"。铁城嶂为火山喷溢形成的流纹岩,经断裂切割,形成峭壁、岩嶂,是雁荡山深谷与叠嶂组合的典型。岩壁上部流动构造十分清楚,下部岩石有所不同,易剥落形成洞穴,水帘洞为其代表。

Tiecheng Peak

Tiecheng Peak of Jingming Valley is located in the Sanzhepu Scenic Area. Tiecheng Peak stands "upright like the Great Wall of China and its colour is the same as black iron". Among the smaller peaks, it is particularly outstanding and has been crowned as "the number one peak of Yandangshan". It was composed of rhyolite formed during volcanic effusion and later being rifted by faulting to create cliffs, screens, gorges and overlapping peaks. The flow structure of rhyolite is very clear in the upper section of the cliff which differs very much from the lower section where the rocks are weathered to form grottos.

净名谷

净名谷,又名初月谷,幽静之至,被称为"净园",雁荡山国家森林公园就坐落于此。 此处的涧谷是由断裂切割巨厚的流纹岩层,随着地壳的抬升,流水的侵蚀而逐渐形成的。

Jingming Valley

It is also known as Chuyue Valley and has been called the "Garden of Tranquillity". It is also where the Yandanshan National Forest Park is located. The valley and gorge were formed by faulting, uplifting and water erosion on the gigantic and thick rhyolitic layers.

铁城嶂 Tiecheng Peak

净名谷 Jingming Valley

展旗峰

展旗峰与天柱峰相对,高260米,状如展开的旗帜,故此得名。此处发育典型的流纹岩层,记录了距今约1亿年前熔岩流动的痕迹。清代著名诗人袁枚(1716—1797)以黄帝战蚩尤的故事进行引申得诗:"黄帝擒蚩尤,旌旗不复收。化为石步障,幅幅生清秋。"

Flying Flag Peak

With a height of 260m and being opposite to Tianzhu Peak, this peak looks like an opened flag. The rocks are rhyolites with very typical flow-banded structure which are strong evidences of lava flow about 100 million years ago. Yuan Mei(1716-1797), a famous poet in Qing Dynasty, expounded based on the battle between Emperor Huang and Chiyou and wrote in his poem that "After Emperor Huang defeated Chiyou, flags flew in the sky and ever since then the flag transformed to become rock curtains and every curtain was magnificent and overwhelming".

展旗峰　Flying Flag Peak

天柱峰

　　天柱峰峰高270米，宽250米。清代喻长霖有楹联曰"左展旗，右天柱，后屏霞，数千仞，神工鬼斧，灵岩胜景叹无双"。其形态由流纹岩受断裂风化剥落塑造而成。

Tianzhu Peak

　　This peak is 270m high and 250m wide. It was described by the famous poet Yu Changlin of Qing Dynasty as "Flying flag on the left, Tianzhu Peak on the right, glow at the back, several thousand yards high, extraordinary craftmanship. The scenery of Lingyan is second to none". The spectacular shape was formed by faulting, weathering and erosion on rhyolitic peaks.

小龙湫

　　小龙湫位于灵岩景区深处，瀑布隐于幽谷之中，瀑水从断崖飞流而下，高70余米。水触石腾空，瀑下有潭，潭外有流纹岩倒石堆积的涧溪。一块巨石如砚，与卓笔峰、卷图峰组成"文房四宝"景观。这里的岩块崩落和流水冲刷现象引发了宋代著名政治家、科学家沈括提出流水侵蚀学说思想。

Little Dragon Waterfall

　　The waterfall is hidden inside the valley of Lingyan Scenic Area. It is about 70m high and when falling water hits the rock, it splashes high up to the sky. There is a plunge pool underneath the waterfall with fallen rhyolitic blocks scattered and piled up outside and along the stream. One of the gigantic blocks looks like an ink stone, lying together with the Zhuobi（brush pen）Peak and Juantu（rolled drawing）Peak to form the "Four Treasures of Study"（Chinese writing tools）. Shen Kuo, the famous politician and scientist of Song Dynasty, had developed a fluvial erosion theory on the basis of phenomena.

天柱峰　Tianzhu Peak

小龙湫　Little Dragon Waterfall

方洞火山岩剖面

方洞凝灰岩位于方洞景区停车场西侧崖壁之上。这里是雁荡山第三期火山活动遗迹的经典观察点，陡壁上的岩层明显地被中部的"金腰带"分为3层。上部为火山爆发时喷溢到空中的火山灰降落到地面堆积而形成的；中部的"金腰带"为火山溢流出来的岩浆形成的流纹岩；下部为火山爆发时沿地表漫延的火山碎屑流堆积而成的。

Fangdong Volcanic Rock Section

The tuffs in Fangdong can be seen up in the cliff to the western side of the parking lot. This is the best observation spot of the 3rd stage of Yandangshan's volcanic eruption. The three tuff strata were obviously separated by a "Golden Belt". The upper stratum was formed by the settlement and accumulation of volcanic ash being thrown out into the sky during eruption. The middle stratum or the "Golden Belt" was composed of rhyolite formed by lava effusion. The lower stratum was formed by pyroclastic flow running on land surface during volcanic eruption.

方洞火山岩剖面　Fangdong Volcanic Rock Section

方洞栈道

方洞栈道是方洞景区的主要游步道，沿1100余米长的栈道而行，你既可以俯瞰周边的自然美景，又可探究岩层的形成缘由和接触关系。沿栈道依次有方洞（又名慧性洞）、关刀洞、云天桥、金钟罩、花瓶洞、凌霄阁、乐仙阁、梅花洞和聚仙阁等景观。游步道深处有铁索桥，长108米，离谷底192米，登桥览胜有惊无险，是一种奇妙的体验。

Fangdong Plank Road

The 1100m long plank road is the primary walking path in Fangdong Scenic Area. Tourists are able to appreciate the stunning scenery as well as exploring the volcanic rock formation along the way. The road passes through interesting geosites such as Fangdong Cave (also known as Huixing Cave), Guandao Cave, Yuntian Bridge, Golden Bell, Huaping Cave, Lingxiao Pavilion, Yuexian Pavilion, Meihua Cave and Juxian Pavilion. At the end of the Plank Road, there is a cable bridge of 108m long which is 192m above the valley bottom, providing unforgettable, thrilling and breathtaking experiences for tourists.

方洞栈道（又称云崖天廊） Fangdong Plank Road（also known as Yunya Gallery）

地质遗迹景观
Geological Relic Landscape

金带嶂

金带嶂位于方洞与仰天湖之间,从上灵岩村、下灵岩村远望最佳。高300余米的岩嶂中间,横贯一条金黄色的近水平岩层——金腰带。沿此岩层开凿栈道,一路奇岩异洞,俯仰绝壁,远眺层峦,铁桥横跨,构成缥缈神奇的山水画卷。

从地质方面来讲,这里揭露了火山在不同时期喷出的不同物质成分,形成的堆积物在纵向上存在差异的现象,此处是科学价值与美学价值的完美结合。1996年第三十届国际地质大会国内外地学专家曾到此考察。

Golden Belt Cliff

Golden Belt Cliff is located between Fangdong Cave and Yangtian Lake. It can be best viewed from Shanglingyan to Xialingyan. It is a golden belt of rock stratum 300m above the ground which extends horizontally among the rock strata. By walking along this plank road, one can capture all the fascinating picturesque scenery of mountains, cliffs and hanging bridge of the area.

Geologically, the rocks have revealed the different materials produced at the different stages of volcanic eruption, causing the vertical variations of the strata in colour, shape and compositions. This also demonstrates the integration of the aesthetic and scientific values when viewing the landform of Yandangshan. Therefore the geopark has been chosen as a field visit site for domestic and overseas geologists of the 30th International Geological Congress in 1996.

金带嶂　Golden Belt Cliff

观音峰

观音峰位于上灵岩村北，观音峰海拔910米，状如观音坐于莲台之上。从方洞入口处或上灵岩村观赏最佳。

观音峰由下至上反映了雁荡山火山第二、第三、第四期火山活动的形成过程，是一处经典垂直剖面。"莲座"以下为雁荡山第二期火山喷发的熔岩形成的巨厚层流纹岩；"莲座"为第三期爆发形成的凝灰岩，其中夹有一层厚2～10米的流纹岩。"观音座身"为第四期火山爆发形成的熔结凝灰岩。

Guanyin Peak

Guanyin Peak is located to the north of Shanglingyan Village. It has a height of 910m and the shape looks like the Goddess of Mercy sitting on a sacred lotus flower. The best viewing point is at the entrance of Fangdong Cave or in Shanglingyan Village.

From bottom to top, the cross section of Guanyin Peak clearly illustrates three different stages of volcanic activities in Yandangshan. Below the so-called "lotus" is a thick rhyolite layer formed by lava effusion during the 2nd stage of volcanic activities of Yandangshan. The "lotus" itself is volcanic tuff formed during the 3rd stage of volcanic eruption, which is sandwiched by a layer of 2–10m thick rhyolites. The "Guanyin" is made of welded tuff formed during the 4th eruption of the volcanic eruption.

观音峰　Guanyin Peak

百岗尖日出　Sunrise in Baigangjian Mountain

流纹构造

雁荡山园区到处可见清晰的流纹构造，它不是沉积作用形成的层理，有人说雁荡山曾沉没大海，这是有误的。这种纹理是火山喷溢出熔岩在地表流动留下的痕迹，称为流纹构造。部分地区的流纹岩里分布的许多"石球"引起了人们的注意，它是含有气体的岩浆溢出地表后，在流动过程中气体局部聚集，形成有空腔的球状岩石。

Flow-Banded Structure

Flow-banded structure of rhyolite can be seen everywhere in Yandangshan Scenic District. These flow bands are not the result of sedimentation but the marks of lava flow left by volcanic effusion. They can be normally found in lava flow. Some of them carry "rock balls", which were formed by gas trapped inside lava during flowing, and were later cooled and solidified to form "balls" with empty centre but very hard shell.

流纹构造　Flow-Banded Structure

球泡流纹岩　Lithophysal Rhyolite

剪刀峰

剪刀峰位于大龙湫景区,是景区的核心景观之一。沿锦溪进入景区,首先看到一座耸立的岩峰。峰上部一分为二,状如指向蓝天的剪刀,故称剪刀峰。清代钱宾王说"百二峰形名不同,此峰变幻更无穷"。此峰是体验移步换景的典型。绕峰而行依次出现"剪刀""啄木鸟""熊岩""一帆"和"桅杆"等造型景观,故一峰多名。

剪刀峰岩石为火山喷溢的流纹岩,经过断裂切割,周围岩石崩落成为孤峰。岩峰发育节理(裂缝)使岩峰中部裂开。从不同侧面观之有不同的造型,极具观赏价值。

Scissors Peak

Scissors Peak is one of the main attractions of Dalongqiu Scenic Area. Walking along Jinxi Stream into the scenic area, the first feature to see is a tower standing right in front with its top part separated into two and pointing to the sky likes scissors. Qian Binwang, a poet in Qing Dynasty said, "a peak may have different shapes and being called different names. But the shapes of this peak are ever-changing and make it hard to have a correct name." It is the typical example of experiencing moving scenery as the scenery continually changes when one moves. It could be like a pair of scissors, woodpecker, bear, sail, mast and so forth. There are so many different names but they all refer to the same peak.

Scissors Peak is composed of rhyolites formed after volcanic eruption. Faulting, weathering and erosion caused disintegration to form individual hills and peaks. The scissors were actually separated by a vertical joint (or crack) right in the middle. It has high aesthetic value and is one of the favourite spots of tourists in Yandangshan Scenic District.

剪刀峰 Scissors Peak

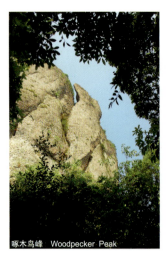
啄木鸟峰 Woodpecker Peak

桅杆峰 Mast Peak

熊岩 Bear Rock

移步换景——剪刀峰 Scenery moves as you move—Scissors Peak

火山通道

火山通道位于大龙湫景区，剪刀峰西北。在锦溪边观景平台，岩壁上似有"龙"字天书，任人想象。其实，这里在地质学上具有重要意义。火山喷发的熔岩流动大多是近于水平的，而此处的流动纹理却从下而上，由直立到弧形，这表明岩浆很可能是通过这里挤出地表，大龙湫的流纹岩很可能从这里喷溢出的。

Volcanic Vent

The volcanic vent is located in the northwest of Scissors Peak in the Dalongqiu Scenic Area. There is a sightseeing platform beside the Jinxi Stream and on the wall of the cliff there is the Chinese character "龙" (dragon). The vertical flow banding and the visible dome shape structure of the original magma chamber indicate that this is the exact location of the volcanic vent where the original magma was extruded and is therefore of great geological significance. It is also the origin of the rhyolite found in Giant Dalongqin.

火山通道　Volcanic Vent

大龙湫

大龙湫位于大龙湫景区最深处,锦溪分支尽头,龙湫背下。大龙湫是雁荡山代表性景观之一,瀑高197米,是我国四大名瀑之一。相传唐初,开山祖师诺讵那进山建寺,后坐化于大龙湫。瀑左有忘归亭,西坡有龙壑轩,岩壁有摩崖石刻20多处。瀑布从龙湫背直泻而下,康有为曰:"一峰拔地起,有水自天来。"徐霞客于1613年3月登上龙湫背和雁湖考察,查明大龙湫水的源头不是来自雁湖,纠正了史书上有误的记载。

Giant Dragon Waterfall

Giant Dragon Waterfall is located at the end of the tributary of Jinxi Stream under the Longqiubei Mountain in the Dalongqiu Scenic Area. It is an iconic landscape of Yandangshan Scenic District with a height of 197m which is one of the four most famous waterfalls in China. As early as Tang Dynasty, Nuojunuo arrived at the mountain and built a temple. He passed away while sitting cross-legged in Dalongqiu. There is Wanggui Pavilion standing on the left of the waterfall and Longhe Pavilion on the western slope. More than 20 inscriptions are found on the nearby precipices. The water originates from Longqiubei Mountain. Kang Youwei, a politician in Qing Dynasty, once described it as "A peak rises from the ground and the water comes from the sky". In March of 1613, Xu Xiake hiked to the top of Longqiubei Mountain and visited the Yanhu Lake. He concluded that Yanhu Lake was not the water source of Dalongqiu Waterfall and corrected the wrong records in previous references.

大龙湫　Giant Dragon Waterfall

西石梁洞

西石梁洞位于含珠峰西1千米处。洞口向南，洞口有大石梁斜倚。洞内有左、右两个洞，右洞长30余米，宽20余米，深约10米，依洞而建木质结构楼房五间。左洞上小下大，上透通而下呈圆锥状，为瀑布冲刷而成。

组成该洞的岩石为流纹岩，从洞口石梁上两组清晰的裂隙（劈理）可以看出，洞是受到北东向和北西向两组断裂带交会切割，岩石破裂、剥落而成。

West Shiliang Cave

West Shiliang Cave is located 1km from the west of Hanzhu Peak. West Shiliang Cave opens to the south and there are big reclining stone beams at the entrance. There are two smaller caves inside, one on the right and the other on the left. The right one is over 30m long, 20m wide and 10m deep. A wooden building with five rooms was built inside. The left cave is narrower in the upper part, and wider and smoother in the lower part, indicating the scouring action of water from the waterfall.

This is a rhyolite cave. Two sets of joints can be seen at the rock near the entrance. They intersected with each other under the influences of a NE-trending fault and a NW-trending fault.

西石梁洞　West Shiliang Cave

西梁大瀑

西梁大瀑位于西石梁洞之西。瀑高约160米,为雁荡山第二大瀑。瀑布沿半个竹筒形的瀑壁直泻到半腰,触石后坠入潭中,气势雄壮。

Xiliang Waterfall

Xiliang Waterfall is located at the west of West Shiliang Cave and is the second largest waterfall of Yandangshan. It has a height of 160m. The cliff alongside the waterfall has a semicircular, bamboo-like shape extending down to the plunge pool.

西梁大瀑 Xiliang Waterfall

芙蓉湖

芙蓉湖位于芙蓉镇东偏南,是雁荡山最大的湖泊,面积约10平方千米。湖中有岛,湖岸曲折,其东南为潮间带滩涂,其北通过黄金溪、芙蓉溪与雁湖山水景观相连。芙蓉湖北靠重峦叠嶂,南对浩瀚东海,是休闲娱乐度假的乐园,堪称为雁荡山的西湖。

Furong Lake

Furong Lake is located at the southeast of Furong Town. It has an area of 10km² and is the largest lake in Yandangshan. There is an island at the centre of the lake. The lakeshore twists and turns with tidal beaches in the southeast. Its northern part passes through two streams called Huangjinxi and Furingxi and connecting the Yanhu Lake. Furong Lake has peaks in the north and faces the East China Sea in the south. It is an ideal holiday resort and has been described as the West Lake of Yandangshan.

芙蓉湖　Furong Lake

雁湖、雁湖岗

雁湖位于雁湖景区北，雁湖岗海拔约800米，古时岗顶有5个湖。雁湖自有文字记载以来已有千年历史，为雁荡山得名之处，有"岗顶有湖，芦苇丛生，结草为荡，秋雁宿之"的记载。

雁湖随岁月而变迁，"雁湖日出"和"雁湖云海"奇观依然充满魅力。雁湖岩石为雁荡山火山喷发结束后，岩浆从地下侵入形成的侵入岩——石英正长岩。

Yanhu Lake and Yanhugang Mountain

Yanhu Lake is located in the northern part of Yanhu Lake Scenic Area beside the 800m high Yanhugang Mountain. There were five small lakes on top of the Yanhugang Mountain in acient times. Historical records of the area can be dated back to a thousand years ago. The name Yandangshan was actually derived from these lakes where the wild geese ("Yan" in Chinese) took them as their habitats.

Although Yanhu Lake changes with time, its sunrise and clouds remain appealing to most people. Its rock is mainly quartz syenite formed by magma intrusion after the end of volcanic eruption.

雁湖　Yanhu Lake

显胜门

显胜门位于显胜门景区、显胜门村西南方。显胜门高200余米，豁口（缺口）顶部相隔仅数米，抬头仰望"自非亭午夜分，不见曦月"，有"天下第一门"之称。门右崖壁上部有著名的石佛洞。此洞为沿裂隙岩块剥落而成。洞内有硅质水溶液淋积成微型柱状体。门内有倒石堆积的"透天十八洞"。门内倒石，证明了显胜门是流纹岩岩峰经断裂、岩块崩落而成的豁口。

Xiansheng Stone Gate

Xiansheng Stone Gate (a natural arch) is located in Xiansheng Scenic Area and southwest of Xianshengmen Village. It has a height of over 200m with a gap of several meters at the top. The sun or the moon cannot be seen from here unless it is at noon or midnight. It has been crowned as "the First Gate under Heaven". The famous Shifo (Stone Buddha) Cave can be found in the upper right side of the cliff. The cave was formed by rockfall along the joints. Inside the cave, minute stalactites are formed by leaching of silica solution. The talus under the stone gate piled to form the famous "Eighteen Caves". The rhyolite blocks prove that the gate was formed by rockfalls along the fault.

生桥,发育在流纹岩岩嶂之上,由于流纹岩层下部富有角砾、球泡而易于剥落,中部发育柱状节理而易破裂崩塌而成。

Xianqiao Bridge

Xianqiao Bridge is located in the northwest of Xianqiao Scenic Area. Looking to the north from Longhu Gate, this natural arch can be seen on the 200m high Xianting Mountain. It has a length of 38m and a width of 8m. Looking down from the bridge, one can have a great view of the surrounding cliffs and clouds which conjure up a fascinating picture of fairyland.

Xianqiao is a typical natural arch found in the cliff of the rhyolitic peaks of Yandangshan. The lower part of the rhyolite layer contains breccia and lithophysa (globular structure) and are subject to quicker weathering and erosion. Columnar joints can be found in the middle part of the peak.

显胜门　Xiansheng Stone Gate

仙桥

仙桥位于仙桥景区西北方向。由龙虎门北望,海拔200余米的仙亭山上有拱形石桥,又名天生桥,桥长38米,宽8米。登桥下俯绝壁万仞,云行流水,如入飘飘欲仙之境。

仙桥为雁荡山流纹岩发育最完美、最典型的天

仙桥　Xianqiao Bridge

龙湖

龙湖位于福溪下游。十里长湖,湖岸曲折,水流如龙。两岸山峦叠翠,奇峰耸立,柱石林立,美不胜收。湖左岸山体由五边形、六边形石柱排列成石柱山。石柱是雁荡山火山第四期(距今1.17亿年)喷发形成的熔结凝灰岩,由高温火山碎屑流均匀冷却收缩而成。

Dragon Lake（Longhu Lake）

Dragon Lake is located in the downstream of Fuxi Stream. On both sides of the stream, hills, tower peaks and rock columns conjure up a picturesque scenery. The polygonal and hexagonal columns on the left of the lake formed a columnar peak. These columns were products of the fourth stage of volcanic activities at Yandangshan (about 117 million years ago). They are mainly welded ash tuffs formed by pyroclastic flows of extremely high temperature followed by even contraction during cooling.

龙湖　Dragon Lake

地质遗迹景观
Geological Relic Landscape

仙姑洞

仙姑洞位于仙桥景区龙湖北。一峰凸起，峰腰悬一洞，是一个呈曲尺状奇洞。洞高约20米，宽30多米，深近40米，有东、南两个洞口。洞口临绝壁，入洞须缘梯攀升。洞顶呈穹隆状，洞内明亮而幽深，洞外层峦叠嶂而高旷。仙姑洞的岩石为雁荡山火山第二期喷溢的流纹岩，洞的形成与两组相交的断裂有关。

Xiangu Cave

Xiangu Cave is located in the north of Dragon Lake in Xianqiao Scenic Area. It can be found in the middle level of the mountain. The cave is 20m tall, 30m wide and 40m deep. It has two entrances, one in the east and the other in the south. As the cave entrance is close to the cliff, one needs to go up the stairs before reaching the entrance. The cave is concave in shape and has sufficient lightings inside. Looking outside from the cave, we can have a panoramic view of the outside peaks. The rock of Xiangu Cave is rhyolite which belonged to the lava erupted at the 2nd stage of volcanic eruption in Yandangshan. The formation of the cave was closely associated with the intersection of two different sets of faults.

定海峰

定海峰位于羊角洞景区，又名剑岩、剑峰。它为一座50米高的孤立岩峰，由流纹质凝灰岩构成，四面绝壁挺拔，远观似剑，经风化崩塌分离而独立成峰。

Dinghai Peak

Dinghai Peak, having been called Jianyan Rock, is located in Yangjiaodong Scenic Area. It is a 50m high isolated rock peak of rhyolitic tuff. It has tall and steep cliffs on its four sides and looks like a sword from distance. Weathering and erosion were the main causes for its present shape.

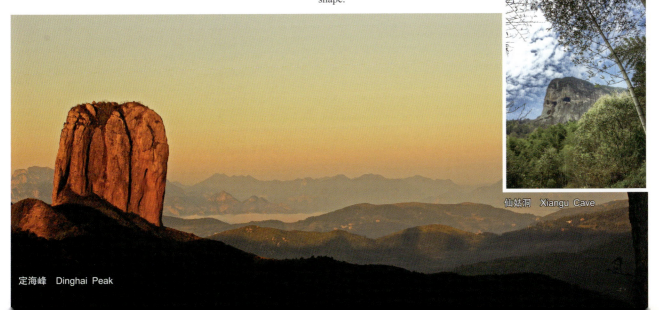

仙姑洞　Xiangu Cave

定海峰　Dinghai Peak

羊角洞

羊角洞位于方山西南侧，万象嶂悬崖。它由劈理带内破碎的岩石崩塌而形成，洞下部宽大达数米，上部有一条劈理缝，因状如羊角而得名。洞内有裂隙泉，称羊角水，洞口建道观玉清宫。

Horn Cave

Horn Cave is located at the Elephants Cliff in the southwestern part of Fangshan. The cave is formed by detached rocks from the fragmented rocks at the cleavage zone. The lower part of the cave is wider, reaching several metres. The upper part has a joint which looks like a sheep horn. Spring seeps out from the cave and has been regarded as horn water. There is a Taoist temple named Yuqing Palace at the cave entrance.

万象嶂

万象嶂为纵览羊角洞景区的南侧山崖。崖高50多米，北西向延伸长600多米，略显曲折，展开成一天然岩屏，又称平霞嶂。因岩壁上直立节理切割形成大量下垂象鼻状景观而得名，从剑岩方向观之，山势更显雄伟。

Elephants Cliff

Elephants Cliff is located at the southern part of Yangjiaodong Scenic Area. It has a height of 50m and extends 600m in a slightly winding way to the northwest to form natural rock screen and is therefore being called Pingxia Cliff. There are a number of sagging elephant trunks like grooves formed by erosion along the vertical joints. If viewed from Jianyan Mountain, the landscape is even more magnificent.

万象嶂 Elephants Cliff

羊角洞 Horn Cave

方山-长屿硐天园区

Fangshan-Changyu Dongtian Scenic District

方山-长屿硐天园区以其优美的自然景观及悠久的石文化著称，位于浙江省台州市温岭市东北部，由方山、长屿硐天两大景区组成。雁荡山火山喷发到空中的火山灰在长屿地区沉降堆积形成了巨厚层的凝灰岩。由于其硬度适中易于开采作为石料，迄今为止拥有1500多年的开采历史，形成了28个硐群，1314个形态各异的硐窟的园区景观，可谓"虽由人作，宛自天开"，堪称"世界第一硐"。除此之外，由水平的流纹岩层构成的方山，四壁皆为陡崖，犹如大地戴上了一尊"礼帽"，道教文化在此盛行，构成了一方宁静的圣土。

Fangshan-Changyu Dongtian Scenic District is famous for its high aesthetic value and very long history of stone culture. It has been called the "world's number one cave" by visitors. The scenic district is located in Taizhou, northeast of Wenling City. It has two major scenic areas: Fangshan and Changyu. The volcanic eruption of the Yandangshan produced a very thick layer of tuff covering the area. Because of its moderate hardness, it was relatively easy to quarry for building materials since 1500 years ago. The cave system has 28 groups with 1314 smaller caves of different shapes. The Changyu Cave has often been regarded as "a naturally created man-made cave". In addition, the cliffs and walls formed by the horizontally layered rhyolite of Fangshan resemble a Taoist's ceremonial hat which has added the holiness of this sacred area.

 镇山嶂

镇山嶂是方山西侧悬崖的总体景观，悬崖高50多米，蜿蜒曲折地展开，凹处形成幽谷。远观可见雁荡群山之磅礴气势，也可

镇山嶂　Zhenshan Cliff

入幽谷近仰方山绝壁之巍峨。岩壁上有"镇山"两个大字而得名镇山嶂。方山四面悬崖形成的各种嶂状景观是熔岩台地地貌的基本表现。

Zhenshan Cliff

Zhenshan Cliff is located in the western side of Fangshan. It is more than 50m high and twisted to form a deep and secluded valley. From here, you can see Yandangshan and the steep cliffs of Fangshan. Zhenshan Cliff is named for inscriptions "Zhenshan" on rock cliff. The different screen-like landscapes formed in Fangshan are basically lava plateau.

云霄寺

云霄寺建造在方山顶,高耸入云霄而得名。云霄寺始建于清朝嘉庆年间(距今约200年前),因山高风急,历史上曾多次遭受风灾毁坏,2000年重建。云霄寺是方山上历史最久、规模最大的佛教寺院,占地面积3000多平方米,有殿堂楼阁50多间。

Yunxiao Temple

Yunxiao Temple was built on top of Fangshan which means the temple is reaching into the sky. It was initially built during the reign of Emperor Jiaqing of Qing Dynasty (about 200 years ago) and had been destroyed several times by strong wind in history. It was rebuilt in 2000. The temple, having large floor area of over 3000m^2 and being the home to over fifty pagodas and halls, is the oldest and largest Buddhist temple in Fangshan.

云霄寺　Yunxiao Temple

上下天湖

上下天湖位于方山顶西南侧。在方山平顶波缓开阔的凹地上筑坝蓄水形成两个小水库,称为上、下天湖。两个天湖镶嵌在平滑光洁的裸露岩丘之间,清爽明媚,为方山增添了灵气。

Upper and Lower Tianhu Lakes

Upper and Lower Tianhu Lakes are located on the southwestern side on the top of Fangshan. Two small reservoirs are built by damming the water conserved in the depression. The upper reservoir has been called Upper Lake and the lower one has been called Lower Lake. These two water bodies are surrounded by barren rocks and have added vitality to Fangshan.

凤凰生蛋

凤凰生蛋位于方山顶南天门。沿近东西向劈理发育的刃状尖峰群,在约50平方米的平台上夹持着一已风化成椭球体的数米长的巨石。巨石似蛋,尖峰肖凤而得名。

Phoenix Laying Eggs

Along the east-west trending joints occurred in the sharp knife-like group of pinnacles at the South Gate on top of Fangshan, some gigantic oval shaped boulders can be found on a 50m² rock platform. These circular boulders were a result of spheroidal weathering and often been called Phoenix Laying Eggs.

上下天湖 Upper and Lower Tianhu Lakes

凤凰生蛋 Phoenix Laying Eggs

瑶池与天河

瑶池与天河为相连水体，瑶池位于方山顶与冬瓜背之间。水色碧绿，温润如玉，三面山壁环抱，景色清幽，宛若仙境。瑶池位于情侣峰（文笔峰）、鹊桥、玉女瀑、渡仙桥、冬瓜背、方山顶等景点包围之中。天河是一条沿北西向节理发育的峡谷，长300米，宽30米，深25米，两侧悬崖直立，筑坝蓄水成为悬河奇观。又因河谷如刀切般平直，似王母玉簪划开，因而引发一些美丽的民间传说。天河是沿节理带风化崩塌作用的最终结果，进一步发育将促使平台的分离和孤峰的形成。

Yaochi Pond & Tianhe River

Yaochi Pond and Tianhe River are two linked water bodies. Yaochi Pond is located between Fangshanding and Dongguabei. The aqua-green and crystal clear water surrounded by mountains on three sides makes an excellent scenery. Yaochi Pond is surrounded by Lover's Peak（Wenbi Peak）, Queqiao Bridge, Yunv Waterfall, Duxian Bridge, Dongguabei and Fangshanding. Tianhe is a valley developed along the NW trending joint with a length of 300m long, a width of 30m and a depth 25m. Cliffs on two sides are steep. The water dam creates a view of suspended river. As the valley is straight and looks being cut by a knife or the jade hairpin of the Goddess of Heaven, therefore beautiful folk legends were created here. Tianhe River was formed by weathering and collapse along joints. Further development will cause separation of platform and formation of isolated peaks.

天河　Tianhe River

瑶池　Yaochi Pond

方岩书院

方岩书院坐落于梅雨瀑与白龙瀑之间,始建于明弘治二年(公元1489年),重建于2008年。整个书院占地1700多平方米,建筑面积2010平方米。书院内设有美丽家园馆、东瓯古国馆、大溪圣贤馆三个部分,运用多媒体全方位展示大溪地区的自然风光、文化历史。

Fangyan Academy

Fangyan Academy, located between Meiyu Waterfall and Bailong Waterfall, was founded in the second year of Hongzhi's reign in Ming Dynasty (A.D.1489). It was rebuilt in 2008. The whole academy has a floor area of over 1700m^2 and the total building area is 2010m^2. There are Halls of Beautiful Home, Historical Dongou and Daxi Sages. It also displays the natural environment and cultural history of Daxi area by applying multi-media technology.

白龙瀑

白龙瀑位于方山悬崖西侧。白龙峡冲沟下端的瀑布,为方山八瀑中最大者。瀑布自80米高的峭壁下坠,颇为壮观。瀑下白龙潭,石为底,水浅见底,泄出成小溪。潭广约50平方米。

Bailong Waterfall

Bailong Waterfall is located on the west side of Fangshan cliff. The waterfall is found at the lower end of the Bailong gully. Fangshan has a total of eight waterfalls and Bailong Waterfall is the largest of all. The water drops down from the 80m cliff, creating a magnificent scenery. Bailong Pond under the waterfall has shallow crystal-clear water which flows out and forms a stream. The pond has an area of about 50m^2.

方岩书院　Fangyan Academy

白龙瀑　Bailong Waterfall

水云硐群

水云硐群位于长屿硐天景区,由52个硐体组成,总面积约1.5万平方米,大部分为现代开采的大型覆钟式采石遗址。其中的中国石文化博物馆为我国最大的硐穴博物馆。《神雕侠侣》《鹿鼎记》等多部电视剧剧组曾取景于此,为水云洞增添了亮丽色彩。

Shuiyun Caves

Shuiyun Caves located in Changyu Dongtian Scenic Area, Shuiyun Caves consist of 52 caves and have a total area of about 15,000m^2; most are large inverted-bell modern quarrying sites. China Stone Culture Museum in Shuiyun Caves is the largest cave museum in China. The famous TV series such as *Condor Heroes* and *Royal Tramp* took shooting here, which adds extra attraction to Shuiyun Caves.

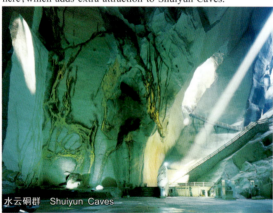

水云硐群 Shuiyun Caves

采石现场 Quarry Site

水云硐群 Shuiyun Caves

地质遗迹景观
Geological Relic Landscape

观夕硐群

观夕硐由308个硐体组成,面积达5.38万平方米,硐体高低错落,彼此连通,最高处离地百余米,完整地保存了隋唐(距今约1500年)至今各时期的采石遗址,是采石遗址宏观形态的最佳展示区。有九曲桥、岩硐音乐厅、硐天宝碗、观音石刻等景点。

Guanxi Caves

Guanxi Cave (Sunset Viewing Cave) has an area of $53,800m^2$. It has 308 caves of different sizes which are interconnected with each other. The tallest cave is about 100m. These caves are relics of Sui and Tang Dynasties. The main attractions are Bridge of Nine Turnings, Cave Concert Hall, Treasure Bowl, and Guanyin Sculpture.

泉声石韵　Spring Water and Rock

龙柱　Dragon Carving Pillars

观夕硐群　Guanxi Caves

硐天花园　Dongtian Garden

岩硐音乐厅

　　岩硐音乐厅位于长屿硐天景区观夕硐内,是典型的大型覆钟式采石遗址。岩硐音乐厅高约70米,直径约100米,总面积2000平方米,能同时容纳近700人入座观赏,具有良好的回声与音响效果。无需电器设备,全靠原声民族乐器,你就可以站在任意一个角落欣赏到自然的立体声了。2002年,德国的北莱州交响乐团还在此举行了一场"莱茵河之声"岩硐音乐会,真是别开生面!

Cave Concert Hall

　　This interesting concert hall is located in Guanxi Cave of Changyu Dongtian Scenic Area. It is a large inverted-bell like quarrying site and has a height of 70m and a diameter of 100m. The total space of the hall is 2000m^2 and can easily accommodate 700 audiences. The echo and sound effects are superb despite no electrical acoustic equipment are used. Normally, folk instruments are played to make use of the natural sound reflected by the walls of the hall. In 2002, the Nordrhein Symphony Orchestra of Germany organised a cave concert called the "Sound of Rhine" in this hall.

岩硐音乐厅　Cave Concert Hall

地质遗迹景观
Geological Relic Landscape

石窗迷宫

石窗迷宫选用了210块当地民间收集的古代石窗,围绕"圆心"排成一个方阵迷宫,置身其中,既可体验走迷宫的乐趣,又能欣赏明、清、民国等不同时期的建筑风格。

石窗之妙在于既实用又有艺术感,还寄托了人们美好的愿望,仔细观察石窗的图案各不相同,它们均是由不同的物体或者字变形而来,代表着相应的寓意。

Labyrinth of Stone Windows

Labyrinth of Stone Windows consists of 210 historical stone windows forming a phalanx labyrinth around "the centre of a circle". By walking through the labyrinth, one can view and appreciate the architectural styles of Ming, Qing and the Republic of China.

Stone windows are both practical and artistic. They also have important cultural implications and historical value. The different patterns represent the different expectation and wishes of the residents living in the houses hundreds of years ago.

石窗迷宫 Labyrinth of Stone Windows

楠溪江园区

楠溪江位于永嘉县境内，以典型的流纹质火山岩地貌、秀美的水体景观和历史悠久的古村落群为主体，具有"水美、岩奇、瀑多、林秀、村古"的特色，是中国耕读文化与山水文化完美结合的杰作。这里发育的白垩纪（距今约1亿年）流纹质火山岩经地质作用不仅形成了崇山峻岭、险壑幽谷，还孕育了楠溪江深厚的文化底蕴，王羲之、谢灵运等历史人物都曾担任过永嘉太守，对传承千年的耕读之风产生了巨大的影响。永嘉学派、永嘉四灵、永嘉昆曲，在中国文化史上具有重要地位。

Nanxijiang Scenic District

Nanxijiang Scenic District is located in Yongjia County. The area is dominated by rhyolitic landform, attractive water features and old villages. It is a perfect integration of traditional Chinese farm-reading and landscape culture. The Cretaceous rhyolitic volcanic rocks, through geologic processes, have not only formed lofty, precipitous mountain ranges, steep and serene valleys, but have also cultivated profound cultural connotations. Historical figures such as Wang Xizhi and Xie Lingyun once served as prefecture chiefs of Yongjia County, had greatly influenced the farm-reading culture that has existed for thousands of years. The Yongjia School, Yongjia Four Poets and Yongjia Kun Opera all played significant roles throughout the history.

楠溪江 Nanxijiang

石桅岩

"沿溪寻胜迹,迎来石桅岣。独出群峰间,屹立惊乾坤。"石桅岩三面临水,奇峰拔地而起,相对高度306米,故被誉为"浙南天柱"。因其形似船桅,故名曰"石桅岩"。

约1亿年前,火山喷出的岩浆冷却后形成火山岩,厚度巨大的火山岩受断裂影响,加之流水侵蚀导致岩石不断崩落后退,最终造就了石桅岩的雄伟气势和独特的地质地貌景观。

Shiwei Peak

"Seeking historical sites along the stream and coming across the Shiwei Peak. It is unique and different from other peaks and had surprised the universe." Shiwei Peak, facing water on three sides, rises up to a relative height of 306m, has been named "Tianzhu of South Zhejiang". It has the shape of a boat mast.

About 100 million years ago, lava from volcano cooled to form rocks. The super thick lava was then subject to faulting and fluvial erosion which eventually led to rockfall and formation of the unique landscape of Shiwei Peak.

石桅岩　Shiwei Peak

仙人造田

岩石上有许多突起的条纹,纵横交错形似乡间田埂。谁能在岩石上犁田呢?恐怕只有天上的仙人了,真不知道是哪位神仙,竟放下此处犁田,只顾自己去潇洒云游了呢,故民间称之为"仙人造田"。从科学的角度讲,它是火山喷发形成的熔结凝灰岩成分不均一,抗风化能力存在差异,抗风化能力弱的凹陷成了"田",反之凸出的成了"埂"。

Divinely Crafted Farmland

The rock has crisscrossing lines on the surface, which looks like traditional Chinese paddy fields. Who would plough on the rock? It would be the immortals from heaven. Which immortal on earth left this ploughed farmland and wandered off? From the scientific perspective, Divinely Crafted Farmland was formed due to heterogeneity and different weather resistance of ignimbrites formed by volcanic eruption. The less resistant part became "fields" and the more resistant parts became "banks".

仙人造田　Divinely Crafted Farmland

灵鹫峰

两座并列的山峰，像一只振翅欲飞的灵鹫，故名。这是雁荡山大规模火山活动形成的火山岩受到断裂切割和长年的风雨侵蚀，周边的岩石风化脱落而形成眼前美景。

Lingjiu Peak

The two peaks stand side by side like a "holy eagle". They were formed by continual weathering and erosion along faults and joints on the volcanic rocks formed by the extensive volcanic activities in Yandangshan.

武士岩

在远处的绝壁上，有两处凹槽，与整个山体组成了眼窝和头部，像一个戴着头盔的古代武士。凹槽部分是岩石受差异风化影响，风化掉了相对脆弱的岩石而形成的。

Warrior Cliff

The two grooves on top of the cliff look like the eyes and head of an ancient warrior wearing a helmet. They were formed by the removal of weathered materials on the relatively weaker parts of the rocks.

灵鹫峰 Lingjiu Peak

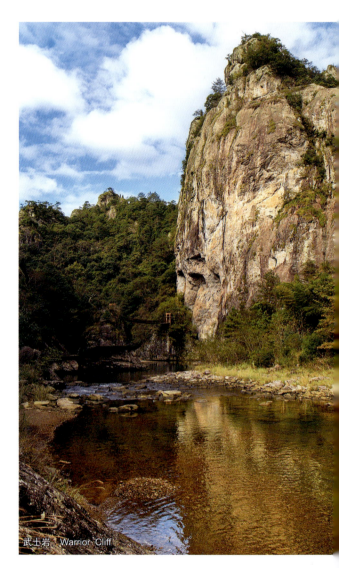

武士岩 Warrior Cliff

麒麟峰

　　两峰耸立,形似传说中的神兽麒麟。这是断层造成的影响,由于破碎的岩石容易风化剥落,而没有破碎的岩石则不易风化,突兀出来。该条断层与形成水仙洞的断层为同一条断层,从水仙洞向麒麟峰望去,你所观察的方向即为断层的方向。

Qilin Peak

　　The two vertical peaks look like mythical unicorns. The shattered rocks caused by faulting are subjected to faster weathering and erosion. Those parts standing out are relatively harder than the surrounding rocks. The Qilin Peak was formed on the same fault as the nearby Shuixian Cave. The two geosites are in alignment with the same direction of the fault line.

小山峡

　　这里是进入石桅岩景区的天然门户——小山峡,此处峡谷幽幽、两岸群峰叠翠,高不可攀;河水弯弯,碧波荡漾,清澈见底,最深处达10多米,真可谓"舟行碧波上,人在画中游!"石桅岩至岭上约3千米的溪流沿断裂侵蚀,除形成多处"V"字形河谷外,还形成了岩槛、跌水、壶穴、侧蚀槽等丰富多样的河流地貌景观。

Xiaoshanxia（Little）Gorge

　　Xiaoshanxia（Little）Gorge is the natural gateway to Shiweiyan Scenic Area. The stream is crystal clear and is over 10m deep at certain sections. It winds through gorges with inaccessible mountains on both sides. It is picturesque when paddling a boat on the water. The stream from Shiwei Peak to Lingshang Village is about 3km long. Geodiversity is interesting along the stream. The V-shaped valley is formed by erosion along a fault line. It also features other fluvial geomorphology such as threshold, plunge pools, pot holes and grooves caused by lateral and undercutting by water.

麒麟峰　Qilin Peak

小山峡　Xiaoshanxia（Little）Gorge

人文景点推荐
Recommended Cultural Scenic Spots

章纶墓

明朝礼部尚书章纶的墓地位于雁荡山园区东北部，为乐清市现存规模最大的墓葬之一。墓园坐东朝西，由墓室和四坛组成，像大多数名人墓一样，依山而筑，呈扶椅式，占地面积约500平方米，建于明代宪宗成化二十三年（公元1487年）。1988年，新建"纶公章吴同宗诗碑"。1990年，重建石碑坊。

Zhanglun Tomb

Zhanglun Tomb is located in the northeast of Yandangshan Scenic District. Zhanglun was the director of Board of Rites in Ming Dynasty. It is one of the largest existing tombs in Yueqing City. The tomb faces west and consists of coffin chamber and four altars with a total floor area of about 500m^2. Like most tombs of celebrities, it was built like sitting on an armchair on a hill. The tomb was built in the twenty-third year of Chenghua's reign in the Ming Dynasty (A.D.1487). The new "poem tablet of genealogy of Zhang-Wu Family" was built in 1988 and the stone monument was rebuilt in 1990.

章纶墓　Zhanglun Tomb

章纶集　The Collected Edition of Zhanglun

章纶故居　Former Residence of Zhanglun

人文景点推荐
Recommended Cultural Scenic Spots

章纶(公元1413—1483年):字大经,号葵心,浙江省温州市乐清市仙溪镇南阁村历史文化名人,明朝正统三年(公元1438年)乡试中举,次年春闱得会魁,殿试中二甲三名进试,正统四年进士。景泰三年,因直言帝位继承一事触犯景泰帝下狱,酷刑拷打,一度濒死,坚贞不屈。三年后英宗复辟,释出,擢礼部侍郎,以立朝刚正著名,后来在礼部侍郎位置上20年得不到升迁。因"性亢直,不能偕俗""好直言,不为当事者所喜"。成化十二年(公元1476年),辞官回乡。去世后被宪宗皇帝追封为南京礼部尚书,赐谥号"恭毅",有《章恭毅公集》存世,后编注为《章纶集》。

Zhanglun(A.D.1413 – 1483): He was also called Dajing with surname Kuixin, was a historical cultural figure of Nange Village in Xianxi Town of Yueqing County in Wenzhou, Zhejiang Province. He passed the imperial examination at the provincial level in the third year of Zhengtong's reign in Ming Dynasty (A.D.1438). He had ranked among the top five candidates in the metropolitan examination the next year. He had then been ranked the third position in the second-class candidates in the final imperial examination and became one of the successfully passed candidates in the imperial examination in the fourth year of Zhengtong's reign. In the third year of Jingtai's reign, he had offended the Emperor by speaking ill of imperial inheritance and was jailed and tortured. However, he remained faithful and persistent. Three years later when Yingzong became emperor, Zhanglun was released and promoted to the position of Minister of Board of Rites and had been well-known for his righteousness in office. However, in the next two decades, he never got promoted due to his behaviours of being "too upright and not sociable in the peer" and "too outspoken and not liked by the others who were in power". In the twelfth year of Chenghua's reign (A.D.1476), he resigned and returned home. After his death, Emperor Xianzong conferred him the title of Director of Board of Rites of Nanjing and a posthumous title "Gongyi" (respectful persistence). He had published the *Collective Articles of Zhang Gongyi* and was later compiled and annotated as *The Collected Edition of Zhanglun*.

章纶像 Statue of Zhanglun

南阁牌坊群

南阁牌坊群位于雁荡山园区北部南阁村的中直街上。牌楼群原有7座,现存5座,皆坐南朝北。5座牌楼沿南阁村主街道一字排列,全长150米。保留较多早期手法,具有明显的浙南地方建筑风格。牌楼立于明正统至嘉靖年间,高7米左右,面阔宽5.5米,进深4米左右。现保存情况较好,周围历史环境未有大的改变。5座牌楼形制、结构相近。均为木石混合结构,单开间三楼悬山式。南阁牌楼群规模宏大,形制完整,是明代牌楼少见实例,既保持了一些官式做法,又有浓厚的地方风格。构造上中柱深埋以稳定整体建筑,空间透露以减轻负荷,具有较高的科学价值。5座牌楼建造时间相隔百年,细微的变化反映了地方风格的演变,具有较高的艺术价值。

Nange Memorial Archway Group

Nange Memorial Archway Group is located in Zhongzhi Street of Nange Village, north of Yandangshan Scenic District. There were seven memorial archways originally with only five still remaining. They are all facing north. The five memorial archways line up along the main street of Nange Village for a distance of 150m. They preserve the early architectural skills of South Zhejiang. They were built in a period from the reign of Emperor Zhengtong of Ming Dynasty to the reign of Emperor Jiaqing of Qing Dynasty. The five archways are similar in shape and structure and are 7m tall, 5.5m wide and 4m thick. They are well-preserved with composite structures of stocks and stones and single-bay three-story overhanging gable roof. Nange Memorial Archways are unique in terms of local styles and characteristics as no other places have similar architectural structures of Ming Dynasty. The middle column was buried deeply to stabilize the whole structure and the space was to reduce load. The long building time of 100 years means minor changes are made to reflect the changes in local styles and preferences. Therefore they have extremely high aesthetic and artistic values.

南阁牌坊群　Nange Memorial Archway Group

灵岩禅寺

　　灵岩禅寺始建于北宋太平兴国四年（公元979年）至今有1000多年历史，为十八古刹之一，因灵岩而得名。禅寺选址讲究，地处嶂壁之间的空地，四周为天柱峰、展旗峰、屏霞嶂等地质遗迹景观。寺外泉瀑流泻，古木参天，林荫夹道，清静幽寂，寰中绝胜。宋朝状元王十朋咏灵岩寺有"雁荡冠天下，灵岩尤绝奇"之句。

　　北宋年间，蒋光赞资助行亮、神昭两位禅师在此建造殿宇达百余间，于是禅风远播，号称"东南首刹"。宋太宗御赐佛经五十卷供奉于寺，真宗赐额"灵岩禅寺"，仁宗赐金字藏经千卷。

　　灵岩禅寺元末毁于兵火，明清年间，多次毁坏，复而重建，直至民国。成圆和尚重修古刹灵岩禅寺，规模相对之前有所缩小，大雄宝殿、天王殿、藏经楼、东西厢房、七如来塔、放生池等焕然一新。

　　近代灵岩禅寺再次被毁，于1984年由显广法师重修恢复，一时灵岩禅寺香火鼎盛，远近十方善信前来敬香礼佛者不计其数，也成为旅游胜地的佛教寺院。

Lingyan Temple

　　Lingyan Temple was built nearly a thousand years ago, in the fifth year of Taipingxingguo Period in the Northern Song Dynasty (A.D.979). It is one of the eighteen historical temples in the area. The site for building the temple was carefully selected. It is located between cliffs and surrounded by multiple geological heritage landscapes including Tianzhu Peak, Flying Flag Peak and Pingxia Cliff. Outside the temple, the presence of springs and waterfalls, ancient trees has created a quiet and secluded environment like heaven. Wang Shipeng, a greatest scholar in Song Dynasty, described the temple as "Yandang is the best in the world while the Lingyan Temple is rare and marvellous".

　　In the Northern Song Dynasty, Jiang Guangzan sponsored two monks, Hangliang and Shenzhao, to build over a hundred of halls here. Since then, Zen's thoughts had spread far and this temple was known as the "First Temple in the Southeast". Emperor Taizong gave the temple 50 volumes of Buddhist scriptures as gift. Emperor Zhenzong named it "Lingyan Temple" and Emperor Renzong presented thousands of volumes of gold-dust-written Tripitaka to the temple.

　　Lingyan Temple was destroyed in a war at the end of Yuan Dynasty. It was destroyed and rebuilt again for several times in the Ming, Qing and up to the Republican period. Later, Monk Chengyuan rebuilt the historical temple to a smaller size, bringing a new look to the Main Hall, Heavenly Kings Hall, Depository of Buddhist Sutras, east and west chambers, Seven Buddhas Tower and Release Pond.

　　It was destroyed again in modern times, and was rebuilt and renovated by Master Xianguang in 1984. From then onward, Lingyan Temple hosts and receives countless worshippers from all over the province and China. It also becomes a famous tourist attraction.

灵岩禅寺　Lingyan Temple

雁荡山摩崖石刻

Yandangshan Inscriptions on Precipices

雁荡山的摩崖碑刻共有370多件,其中唐代的4件,宋代的50件,元代的2件,明代的43件,清代的36件,现代的150多件,以及待考的50多件。灵峰古洞左侧有党和国家领导人江泽民、李鹏的题字与题句数处。

雁荡山摩崖碑刻有题名、题字、题诗、记事、记游等,内容极其丰富多彩,其形式也很多样活泼,题刻文字有长有短,长的多达770字,短的只有一个字;大多刻石只书写文字,也有配上图画的;至于题刻的字体,有楷书、行书、草书,也有篆书、隶书。而且即使同一种字体,也是各具个性特色,风格多样,异彩纷呈。

Yandangshan has more than 370 cliff-side inscriptions in total, among which, four were inscribed in Tang Dynasty, fifty in Song Dynasty, two in Yuan Dynasty, forty-three in Ming Dynasty, thirty-six in Qing Dynasty, over 150 belonging to modern times and another 50 yet to be verified. On the left side of Lingfeng Ancient Cave, there are several inscriptions of ex-president Jiang Zemin and ex-premier Li Peng of modern China.

Yandangshan cliff-side inscriptions are engraved with colourful and meaningful contents in different types. These include autographs, characters, poems, anecdotes and travel notes. The inscriptions can be long or short. Some may be as long as 770 characters or short as only one character. Most inscriptions are in form of characters but some have pictures. The writing styles can be regular script, running script, cursive script as well as seal and official scripts. Each inscription has its own characteristics which make them unique and interesting.

石刻　Stone Inscriptions

人文景点推荐
Recommended Cultural Scenic Spots

岩头古村

岩头古村位于楠溪江园区西南部,是楠溪江中游最大的古村落,创建于五代末年,为金氏聚居之地,占地18.5公顷(0.185平方千米),平均海拔约49米。因村庄正对"芙蓉三冠"的三岩之首(芙蓉北岩),故名岩头。岩头古名上芙蓉,据传祖先于唐末清泰年间,从福建长溪,迁居永嘉档溪西巷,后羡芙蓉三岩胜境而再迁与此。宋末元初,始祖金安福(公元1250—1318年)迁来此地。到明世宗嘉靖年间(公元1522—1566年),由八世祖金永朴主持,进行全面规划修建,平面上略呈矩形。

岩头村以科学的水利设施和巧妙的村庄布局而闻名,有着楠溪江中游规模最宏大、设计最巧妙的村落供水系统和公共园林。

Yantou Ancient Village

Yantou Ancient Village is located in the southwest of Nanxijiang Scenic District. It is the largest ancient village in the middle reaches of Nanxijiang. It covers an area of 18.5 hectares ($0.185 km^2$) at 49m above sea level. It was founded as the settlement of Jin Family in last years of

岩头古村平面图 Yantou Ancient Village Plan

芙蓉古村　Furong Ancient Village

59

the Five Dynasties. The village is named "Yantou" which means cliff head as it is facing directly to the first cliff of the "Three Rocks of Furong". It was once named as Shangfurong Village. The legend said that the ancestors moved to the Dangxi Western Lane of Yongjia County from Changxi County of Fujian during the years of Emperor Qingtai of late Tang Dynasty and later moved to this area because of the impressive landscape of "Three Rocks of Furong". The earliest ancestor Jin Anfu（A.D.1250－1318) moved here in the late Song Dynasty and early Yuan Dynasty. The eighth generation ancestor Jin Yongpu（A.D.1522－1566) carried out overall planning and construction during years of Emperor Jiajing of Ming Dynasty and the village became the current rectangular shape that we can see nowadays.

Yantou Village is best known for its scientific irrigation design and facilities. It is the largest and the best designed village in the middle reaches of Nanxijiang.

岩头古村——丽水街 Yantou Ancient Village—Lishui Street

苍坡古村

Cangpo Ancient Village

苍坡古村原名苍墩，位于楠溪江园区西北部，西南距岩头古村3千米，始建于五代后周显德二年（公元955年），为李姓聚居之地，占地9.73公顷，平均海拔约50米。现存的苍坡村是九世祖李嵩于南宋淳熙五年（公元1178年）邀请国师李时日按五行风水说，依"文房四宝"布局重建的村落，至今已近840年的历史，平面上略呈方形。

Cangpo Ancient Village, formerly named as Cangdun Village, is located in the northwest of Nanxijiang Scenic District and is 3km away from the Yantou Ancient Village in the southwest. It was founded as a settlement of people after the family name Li in the 2nd year of Emperor Xiande of Later Zhou Dynasty of the Five Dynasties (A.D.955). The village has an area of 9.73 hectares at an average altitude of about 50m above sea level. The ninth generation ancestor Li Song invited the State Preceptor Li Shiri to re-design the village according to the shape of the the "Scholar's Four Treasures". These comprise writing brush, ink stick, ink slab and paper. By adopting the theories of Five Elements and Feng-shui, the reconstruction started in the 5th year of Emperor Chunxi of the Southern Song Dynasty (A.D.1178). Therefore the existing Cangpo Village has a history of nearly 840 years. The setting of the village looks roughly like a square.

苍坡古村平面图　Cangpo Ancient Village Plan

苍坡古村的水月堂
Shuiyue Hall of Cangpo Ancient Village

永嘉昆剧　Yongjia Kun Opera

雁荡山世界地质公园科学故事

Science Stories of Yandangshan UNESCO Global Geopark

雁荡山世界地质公园在地质构造位置上处于环太平洋大陆边缘构造岩浆带中的中国东南沿海中生代火山岩带，浙江省东部是这一火山带中火山岩出露最广的地区，而雁荡山是中国东南沿海火山带最具代表性的古火山遗迹之一。

什么是火山？

火山充满神奇性，虽然从"火山"的英文"volcano"中看不出有"火"的迹象，其实这个单词源自火神的名字——Vulcan。据罗马神话记载，伏尔甘（Vulcan）是人们信奉的火神。他还是造兵器的好手，传说伏尔甘在为众神打造兵器时，专门前往意大利著名的活火山埃特纳火山（Mount Etna），利用火山口涌出的炽热岩浆锻造兵器。从此这位火神与火山结下了不解之缘。而火山的科学说法，是高温的岩浆经过地下通道喷发出地表，由火山喷发物质堆积成的锥状、盾状或圆形或凹陷状地质体称为火山。

Yandangshan UNESCO Global Geopark, in respect to the geologic structure position, is located in the Mesozoic volcanic belt of southeast coastal region of China in the structural magmatic zone of the circum-Pacific continental margin. The eastern part of Zhejiang Province has been regarded as the most extensive area with exposed volcanic rocks of this remarkable volcanic belt. Yandangshan is one of the most typical representations of ancient volcano in the volcanic belt of southeast coastal region of China.

What are Volcanoes?

Volcanoes are full of miracles. However, the English word "volcano" does not carry any sign of "fire". Actually, the name was derived from the name of the God of Fire—Vulcan. According to the Roman mythology, Vulcan is the God of Fire and he was also good at armament manufacturing. The legend told that Vulcan once went to the famous active volcano-Mount Etna in Italy to forge armaments for all gods by using boiling magma of the crater. Vulcan had forged an indissoluble bond with the volcanoes since then. Whereas, the scientific explanation of volcano is that magma of high temperature travels through underground conduits and erupts to the earth surface to form conical, shield, circular or concave geologic bodies with the deposition of erupted materials around the crater.

惊天动地的火山爆发景观

火山爆发是火山喷发的一种强烈的喷发方式。火山爆发的过程是由下而上。

（1）在深处存在溶解挥发成分的岩浆。

（2）岩浆上升后，压力降低，导致挥发出熔浆，熔浆由气泡和液体组成。

（3）岩浆继续上升，压力进一步降低，导致岩浆和岩石成为火山碎屑。

（4）富含火山碎屑的气体形成喷发柱。喷发柱中碎屑物降落堆积后形成岩石，称火山碎屑岩。雁荡山第一期火山喷发，就是这种强烈的火山爆发。所以雁荡山火山一出世就显示其强大的威力。

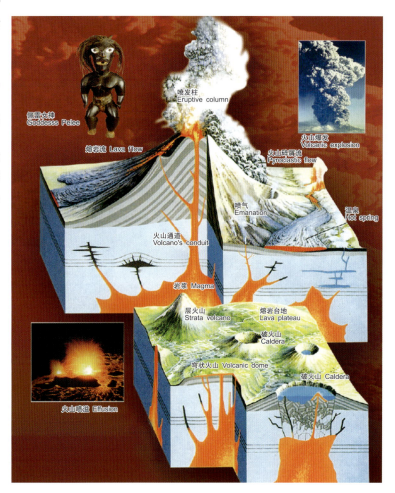

火山家族　Volcano Family

Earth-Shaking Scenery of Volcanic Explosion

Volcanic explosion is a violent mode of volcanic eruption, and it erupts from bottom to top.

（1）Magma with dissolved and volatile components exist deep inside the Earth crust.

（2）As the magma rises, the pressure drops, resulting in trapping of air inside molten magma to form vesicular structure.

（3）With the continual rising of magma, the pressure further reduces. Lava and rock debris start turning into pyroclastic form.

（4）Gases rich in pyroclastic materials could form eruption columns. The pyroclastic materials erupted in the eruption columns fall to ground, being settled, solidified and form rocks. They are called pyroclastic rocks. The 1st stage of volcanic eruption in Yandangshan belonged to this violent mode right at the beginning of the birth of Yandangshan.

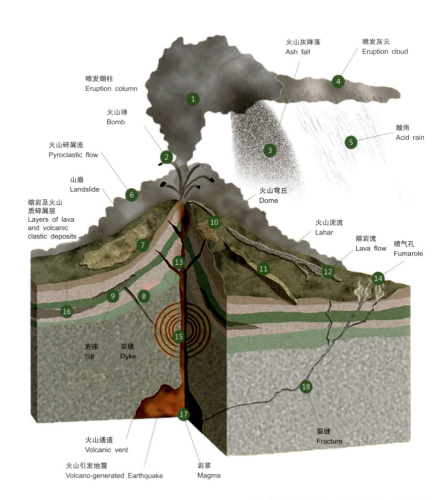

火山爆发模型　Volcanic Explosion Model

火山岩浆喷溢地表的景观

火山喷发的另一种方式称喷溢。它是岩浆从火山口以比较平静的方式溢出地表。熔岩流（河）犹如"火龙"，熔岩湖翻滚，部分岩浆抛到空中犹如一口炼钢的熔炉。雁荡山第二期火山喷出的方式就是这种喷溢。部分岩浆从狭小的通道中挤出，称为岩穹，雁荡山七星洞原为一个火山岩穹。

火墙——裂缝喷发
Magma Wall—Fissure Eruption

熔岩喷泉与熔岩河
Lava Fountain and Lava Rivers

熔岩河
Lava River

熔岩流
Lava Flow

Effusion Spectacle of Volcanic Magma

Effusion is another mode of volcanic eruption. It refers to the process of magma coming out from crater and spilling over the earth surface in a relatively peaceful mode. The lava flows (rivers) look like "fiery dragon" and the lava lakes are just like smelting furnaces, rolling over and throwing some magma into the air. The 2nd stage of volcanic eruption at Yandangshan was in form of effusion. The narrow vent where the magma escaped was known as lava dome. The Qixing Cave was a good example of a lava dome.

Lava Flowed into the Sea and Generated Steam Explosion

Dome-shaped Lava Fountain

Lava Driblet Cone

Lava Tube

Lava Fountain and Lava River

岩石地层柱与同位素年龄——雁荡山火山真实年龄

雁荡山火山具体年龄多大？地质学家按火山先后喷出的一层一层岩石，建立岩石地层柱，然后运用同位素地质学方法测定其年龄。雁荡山第一期喷出岩石（熔结凝岩灰岩，样品取自王家番）年龄为1.28亿年；第二期喷出岩石（流纹岩，样品取自上灵岩到方洞）年龄为1.21亿年；第三、第四期喷出岩石（火山碎屑岩，流纹岩，样品取自方洞到百岗尖）年龄为1.17亿年；最后岩浆侵入石英正长岩（百岗尖）年龄为1.08亿年。1.28亿～1.08亿年，为中生代时期早白垩世。

Lithostratigraphic Column and Isotopic Age—Age of Yandangshan Caldera

How old is the Yandangshan caldera? Geologists have constructed a lithostratigraphic column composed of rock layers successively formed by erupted materials and measured their ages by isotopic dating. Based on this, welded tuffs, sampled at Wangjiafan in the 1st stage of volcanic eruption at Yandangshan were formed 128 million years ago. The rhyolite, sampled from the area between Shanglinyan Village and Fangdong Cave in the 2nd stage of volcanic eruption at Yandangshan dated back to 121 million years ago. The pyroclastic rock and rhyolite sampled from the area between Fangdong Cave and Baigangjian Mountain in the 3rd and 4th stages of volcanic eruption in Yandangshan age were formed 117 million years ago. Besides, the time of magmatic intrusion of quartz-syenites on Baigangjian Mountain was formed 108 million years ago. The period between 128 million years ago and 108 million years ago belonged to the Early Cretaceous in the Mesozoic period.

108 million years ago belonged to the Early Cretaceous in the Mesozoic Era

火山家族

要了解雁荡山的身世，还需要了解火山家族。火山家族中有不同内在性格和外在风貌的火山成员。常见的火山有：

（1）盾状火山底部比较宽，坡度很缓，它是由稀薄的熔岩很快地流动并覆盖了一片相当宽广的地方时形成的。

（2）裂隙式火山是地表上的一条长裂缝，它喷出热而稀薄的岩浆，这种岩浆很快扩散开来，不会形成山丘或山峰，根据四周的地貌，它可能产生一个广阔的火山平原。

（3）有着陡峭山坡（30°~40°）的锥状火山可以高达数百米，它是火山猛烈爆发的结果，爆发时会喷出大股黏稠的熔岩和大块的岩石，这些物质落在火山口附近很快冷却下来，形成圆锥形的碗状火山口。

（4）复合式火山的形成过程，先是一段一段的、安静而持续的小喷发，继之是猛烈的大爆发。随着时间的推移，一层又一层的物质堆积起来，形成陡峭的锥形山坡。

（5）破火山口火山拥有宽而浅的火山口，那是由于火山经历了一次剧烈爆炸造成火山顶部的火山锥部分坍塌而形成的。如果塌陷后发生了更多的火山活动，新的火山锥又会在破火山口里形成并上升形成新的山顶。

火山家族
Volcano Family

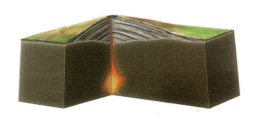

盾状火山
Shield volcano

裂隙式火山
Fissure vent volcano

锥状火山
Cinder cone volcano

Volcano Family

Understanding the origin of Yandangshan involves more knowledge of the volcano family. The volcano family has several members with different characters and features. They include:

(1) A shield volcano is a wide volcano with shallow and gently sloping sides. They are formed by lava flows of low viscosity which allow the lava to flow extensively over large area.

(2) Fissure vents usually form part of the structure of shield volcano. Also known as volcanic fissures or eruption fissures, they are linear volcanic vents through which lava erupts usually without explosion. Their usual width ranges from a few metres to several kilometres. They are fed by dykes connecting deep underground magma chamber and can cause large flood of basalts which run first in lava channels and later in lava tubes.

(3) A cinder cone or scoria cone is a steep conical hill (30° – 40° gradient) built up by loose pyroclastic fragments, such as volcanic clinkers, cinders, volcanic ash and scoria around a volcanic vent. They are loose pyroclastic debris formed by explosive eruptions or lava fountains from a single, typically cylindrical vent. As the gas-charged lava is blown violently into the air, it breaks into small fragments that solidify and fall as either cinders or scoria around the vent to form a symmetrical cone. A cinder cone has a nearly circular ground plan as well as a bowl-shaped crater at the summit.

(4) A composite cone volcano, or a stratovolcano, is built by multiple eruptions from surrounding volcanoes. They are formed over hundreds of thousands of years and have their entire structure built by magma flowing from geographically close volcanoes.

(5) A caldera is a large collapsed volcanic depression. Collapse of a volcanic crater is a result of loss of support due to the emptying of its magma chamber. If volcanic activities prevail, new cones can be formed inside the original depression and become new hilltop in the caldera.

复合式火山
Composite cone volcano

破火山口火山
Caldera

什么是破火山？

破火山是指形态近似圆形的大型火山凹陷,是火山口受到破坏形成的锅形洼地。破火山形成过程比较复杂,一般经历以下几个过程:①火山爆发大量岩浆排出;②岩浆房排空;③火山口周围崩塌下陷形成破火山。雁荡山破火山形成之后火山再次复活,喷溢熔岩并成为岩穹,称之为破火山复活阶段,因此雁荡山火山又叫作复活破火山。

What is a Caldera?

A caldera is a large volcanic depression with a nearly circular shape, in which the crater is a basin shaped low-lying landform created by volcanic eruption. The formation processes involve: ① The volcano erupts and discharges massive magma; ② The magma chamber is vacated; ③ Top part of the volcano collapses to form a large depression called caldera. Yandangshan caldera resurged with new effusion of magma to form lava domes. This is the caldera resurgent stage and therefore Yandangshan caldera is also called a revived caldera.

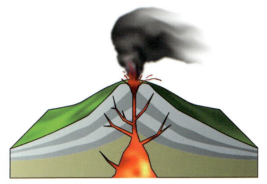

火山爆发大量岩浆排出
The volcano erupts and discharges massive magma.

岩浆房排空
The magma chamber is vacated.

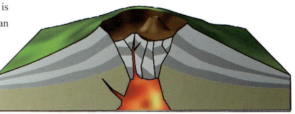

火山口周围崩塌下陷形成破火山
Top part of the volcano collapses to form a large depression called caldera.

破火山形成图解
Caldera Formation Diagram

什么是火山岩？

火山喷发而成的岩石称火山岩。雁荡山火山岩依据其化学与矿物成分可划分为多种类型。雁荡山火山岩石中二氧化硅含量67%～77%，氧化钾与氧化钠含量8%～10%，根据火山岩化学成分分类图，我们将它划为流纹岩类。喷溢的熔岩由于地表冷却条件的差异而形成了各种各样结构的流纹岩。如熔岩流顶部为流纹状流纹岩，中上部为球粒泡流纹岩，中部为斑状、块状流纹岩，下部为球泡流纹岩、角砾流纹岩等。雁荡山世界地质公园的大部分景观均是由流纹岩构成的。

What are Volcanic Rocks?

Rocks generated by volcanic eruption are called volcanic rocks. Volcanic rocks of Yandangshan are divided into many types according to Chemical Classification of Volcanic Rocks. The major type is the rhyolite, containing 67%−77% silicon dioxide and 8%−10% potassium dioxide and sodium dioxide. The rhyolite, with various structures, is formed by effused magma shaped up in different surface cooling conditions. For example, in a lava flow, the top part forms rhyotaxitic rhyolite, the middle−upper part forms spherulite rhyolite, the middle part forms the porphyritic and bulk rhyolite, and the lower part forms the rock globular or lithophysa rhyolite and the breccia rhyolite. Most of the landscapes of Yandangshan UNESCO Global Geopark are formed by rhyolite.

什么是火山岩 What are volcanic rocks?

火山岩化学成分分类图
Chemical Classification of Volcanic Rrocks

蓝方石响岩 Hauynite Phonolite

粗面岩 Trachyte

流纹岩 Rhyolite

玄武岩 Basalt

珍珠流纹岩 Pearl rhyolite

粗面斑岩 Trachyte porphyry

球泡流纹岩 Lithophysal rhyolite

流纹构造 Fluidal structure

球泡空腔 Lithophysa cavity

球泡流纹岩 Rhyolite with globular structure

雁荡山：火山岩的天然博物馆

　　火山碎屑岩是由火山爆发而形成的，由玻璃质碎屑和岩石、矿物碎屑组成。雁荡山的火山碎屑岩种类十分丰富，其化学成分属于流纹质。富含气体的火山碎屑物紧贴地面流动称之为火山碎屑流，形成的岩石即熔结凝灰岩，雁荡山第一、第二、第四期喷发形成此种岩类，构成雁荡山锐峰的也为这一类岩石。

　　凝灰岩是由喷发到空中的火山灰降落到地表固结成岩的，雁荡山第三期爆发的岩石即为凝灰岩，长屿硐天的大部分岩石均由凝灰岩构成，只不过当时的火山灰中夹杂了许多角砾（小岩块），故而称之为角砾凝灰岩，它是极为优质的建筑材料。

Yandangshan: A Natural Museum of Volcanic Rocks

　　Pyroclastic rocks consist of vitric pyroclastic, lapilli and mineral detritus. Yandangshan contains various pyroclastic rocks and have similar chemical compositions as rhyolite. The more volatile pyroclastic materials full of gases flow over the ground is called pyroclastic flow and when they are solidified, welded tuffs are formed. They were formed during the 1st, 2nd and 4th stages of volcanic eruption at Yandangshan. For instance, Ruifeng Peak was made of pyroclastic rocks.

　　Tuffs are formed by volcanic ashes after they are erupted to the air and fall down to the earth surface and are solidified to form rock. Tuffs are formed at the 3rd stage of volcanic eruption in Yandangshan. For instance, Changyu Dongtian are formed of brecciated tuffs as volcanic ashes mixed up with broken fragments during eruption. They are currently used as good quality building materials.

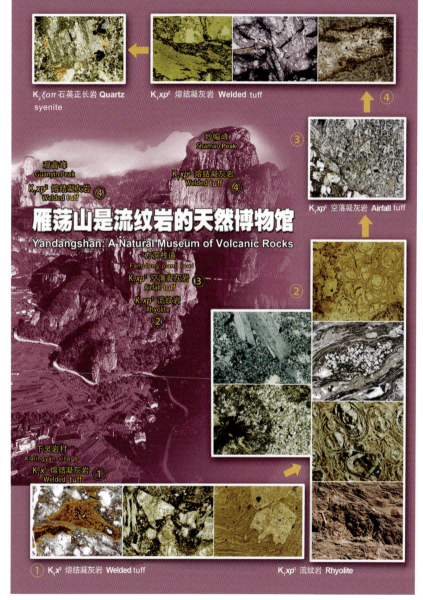

雁荡山火山生命史——雁荡山火山演化模型

Evolution History of Yandangshan Caldera—Evolution Model of Yandangshan Caldera

雁荡山火山活动历史为：

（1）1.28亿年前火山爆发形成大量的火山碎屑流，它们凝固组成了雁荡山第一期火山活动形成的主要岩石类型——熔结凝灰岩。

（2）由于火山爆发，大量岩浆排出，岩浆房腾空，火山上部失去支撑而塌陷，形成了最初的破火山口。

（3）1.21亿年前破火山复活再次喷发，火山岩浆溢流形成巨厚流纹岩层，溢出的岩浆在局部形成了岩穹。

（4）破火山南部爆发，喷发出巨大量的火山灰，这些喷溢的火山灰就近在空中降落，形成空落凝灰岩。

（5）1.17亿年前再次发生大爆发，形成晚期火山碎屑流，进而形成了熔结凝灰岩。

（6）1.08亿年前深部岩浆侵入形成侵入体——石英正长岩。

火山停息后，经过1亿多年地壳抬升与剥蚀，形成如今雁荡山面貌，但仍保留了破火山地貌要素。雁荡山火山的形成与演化模式是亚洲大陆边缘破火山中的杰出代表。

The active history of Yandangshan caldera is as follows:

(1) The first stage of volcanic eruption occurred 128 million years ago and generated massive pyroclastic flows and formed the main welded tuffs after solidification.

(2) Due to the volcanic explosion, a large number of magmas were discharged and the magma chamber was vacated, causing the collapse of the upper part of the volcano because of losing support, thus, the initial caldera formed.

(3) The caldera revived and erupted again 121 million years ago. The effused volcanic magmas formed thick rhyolitic layers and some formed lava domes in certain environments.

(4) The southern part of the caldera erupted and extruded massive volcanic ashes that have formed air-fall tuffs after settling down on the ground.

(5) The caldera exploded again 117 million years ago, creating late pyroclastic flows which formed the welded tuffs.

(6) The magma in deep intruded into rocks 108 million years ago forming intrusive rocks, e.g., quartz-syenites.

The crust experienced uplift and denudation in the past 100 million years after the volcano activities stopped, forming the current appearance of Yandangshan caldera. The Yandangshan caldera is an ideal evolution model and an outstanding representative of caldera formation at the continental margins of Asia.

雁荡山火山第一期爆发形成火山碎屑流
The 1st stage of volcanic eruption generated pyroclastic flows.

破火山塌陷，局部蒸汽岩浆爆发
The caldera collapsed and phreatomagmatic eruption started at some areas triggered pyroclastic flow.

雁荡山火山第二期，破火山复活，火山喷溢，熔岩溢流与侵出
During the 2nd stage of volcanic eruption, the Yandangshan caldera revived and erupted, extruding massive lava to the surrounding areas.

雁荡山火山第三期，火山再次局部爆发，形成空落、火山碎屑流
In the 3rd stage of volcanic eruption, the volcano produced ashfall and pyroclastic flows.

雁荡山火山生命史
Evolution History of Yandangshan Caldera

雁荡山火山第四期，破火山再次爆发形成火山碎屑流
During the 4th stage of volcanic eruption in Yandangshan, the volcano generated extensive pyroclastic flow in the surrounding areas.

破火山再次塌陷，岩浆侵入
The caldera collapsed again followed by magma intrusion.

破火山抬升与剥蚀
The caldera uplifted and erosion started.

雁荡山火山由来

火山活动往往与板块运动有关,火山常分布在板块边缘,在距今约1.4亿年前古太平洋板块欧亚大陆板块俯冲,板块间互相挤压摩擦产生能量,使上地壳与下地壳部分熔融形成岩浆。挤压使地壳破裂形成裂缝,岩浆沿着裂缝上升到地表,火山就喷发了。

The Origin of Yandangshan Caldera

Volcanic activities are often associated with plate movement. Volcanoes are often distributed at the edge of a plate. About 140 million years ago, the energy was generated by extrusion and friction between plates during the subduction of the Paleo-Pacific Plate towards the Eurasia Continent Plate. This caused the melting of the Earth's upper crust and the Earth's lower crust. Compression created extreme pressure and the crust ruptured, allowing magma to extrude to the Earth's surface through cracks to form volcanic eruptions.

雁荡山火山的由来
The Origin of Yandangshan Caldera

雁荡山火山喷发后的地质作用——大自然的力量塑造了雁荡山"不类他山"的风光

雁荡山为什么能呈现"不类他山"的风光？为什么雁荡山自然景观有很高的美学价值？因为雁荡山具有两个前提、六个条件。两个前提：①雁荡山是一座流纹岩浆喷发的大型破火山；②雁荡山处于亚洲大陆边缘海洋性气候环境。六个条件：①雁荡山火山岩层保存良好；②巨厚的流纹岩层是造景的主要"材料"；③断裂及沿断裂的溪涧流水是塑造奇景的神斧；④雨水侵蚀、风化剥落、重力崩塌等作用精雕细刻雁荡风光；⑤植被发育与良好保护装扮了雁荡风光；⑥季节、昼夜、阳光、雨水、风力的变化增添了雁荡风光的变幻性。由此形成了独特的滨海流纹岩山岳生态地学景观。

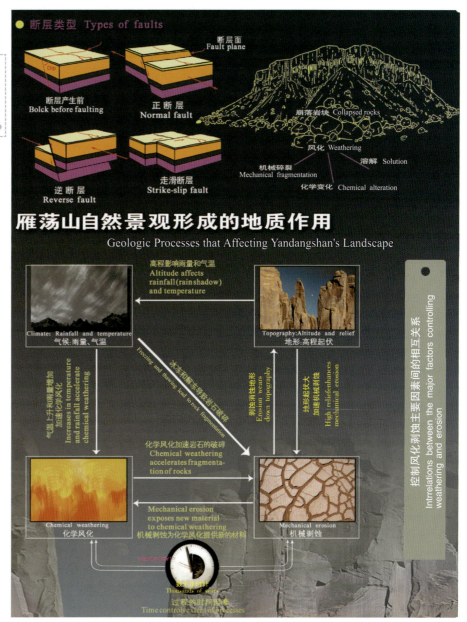

The Post Volcanic Eruption Geological Processes—The Power of Nature Creating the Uniqueness of Yandangshan Scenery

Why does Yandangshan have unique scenery? Why does the natural landscape of Yandangshan have high aesthetic value? Perhaps, the two prerequisites and six conditions would provide the answers. The first prerequisite is that Yandangshan is a large caldera with rhyolitic lava. The second prerequisite is that Yandangshan is located at the edge of the Asian continent with maritime climate. The six conditions are, ①the volcanic rock layer of Yandangshan is well preserved, ②the ultra-thick rhyolite layer is the main "material" for shaping the landscapes, ③faults and joints are keys to the formation of miraculous scenery, ④exogenic factors such as water erosion, weathering and denudation, and gravitational collapse are actively operating, ⑤well-developed vegetation protects and dresses up the scenery of Yandangshan, ⑥seasons, day and night, sunshine, rain, and wind changes add variations to the scenery. These conditions are combined to create the remarkable and unique ecological and geological landscape of coastal rhyolitic mountains.

雁荡山嶂、门、柱的形成

雁荡山自然景观丰富多彩，尤以叠嶂、门阙、独柱较为突出。所谓嶂者，山体悬崖峭壁，方展如屏。它是经断裂切割、岩块滑移、重力崩塌而成的。门阙则是在嶂的基础上经垂直嶂的另一组断裂切割、岩石破裂、岩块崩落，留下两岩对峙如门。独柱则是嶂经多组断裂切割，其周围岩石均经破裂而崩落，留下孤立的柱峰。所以，在叠嶂的基础上可以派生出如"五马回槽""一帆峰""双笋峰""显胜门""南天门""天柱峰"等景观。

Formation of Yandangshan Cliffs, Gates, and Columns

Yandangshan has various natural landscapes, among which, cliffs, gates and columns are extra-ordinary The so-called cliffs refer to the steep mountain cliffs, square and extended like screens. Cliffs are formed by faulting, rock slides and gravitational collapse. Gates refer to two rocks left standing after the cliffs are cut by another group of faults perpendicular to the cliffs. Single columns refer to the isolated column cliffs left after the cliffs were cut by faults resulted in the collapse of surrounding rocks. Therefore, famous landscapes, such as "Five Horses Returning to Shed" "Yifan Cliff" "Double Bamboo Shoots" "Xianshengmen Stone Gate" "Nantian Stone Gate" and "Tianzhu Peak", are all formed along the cliffs.

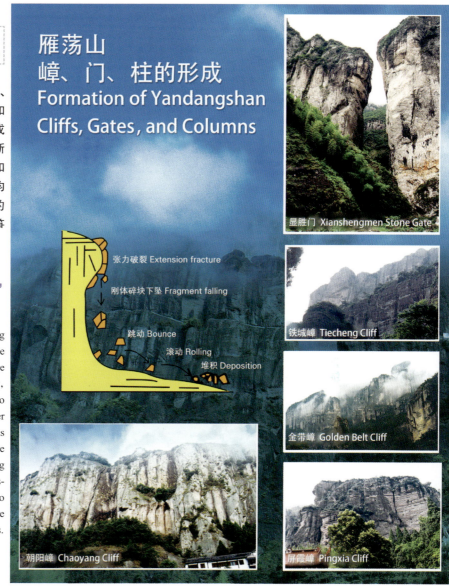

旅游服务
Tourist Services

雁荡山世界地质公园博物馆

雁荡山世界地质公园博物馆占地面积20亩（1亩=666.7平方米），建筑面积3620平方米。分8个展厅和1个展廊，即序厅、影视厅、地质遗迹厅、火山演示厅、文化展示厅、世界火山展示厅、生态展示厅、地质公园建设发展厅及地质展廊。馆内充分展示雁荡山地质地貌特征、地质遗迹特征、火山喷发和破火山形成过程、全球火山特征、生态文化的多样性、地质公园发展历程。博物馆建成至今已接待了多位国家领导人、地质专家、其他地质公园建设管理者及大量的普通游客，也成为了青少年科学考察探险基地、国土资源科普基地、全国科普教育基地。

The Museum of Yandangshan UNESCO Global Geopark

Yandangshan UNESCO Global Geopark Museum covers an area of 20mu (1mu=666.7m^2), with a construction area of 3620m^2. It is divided into eight exhibition halls and an exhibition gallery, namely the Prelude Hall, the Video Hall, the Geological Relics Hall, the Volcanic Demonstration Hall, the Cultural Exhibition Hall, the Global Volcano Exhibition Hall, the Ecological Exhibition Hall, the Construction and Development of Geopark Hall and the Geological Exhibition Gallery. The museum fully displays the geological and landscape features of Yandangshan, characteristics of geological relics, volcanic eruptions and formation process of caldera, characteristics of global volcanoes, diversity of ecological cultures and the development of geopark. Since its establishment, the museum has received many national leaders, geological experts, construction managers of other geoparks and a large number of ordinary tourists, and has become a youth scientific expedition and adventure base, a land and resources science popularization base, and a national science popularization base.

旅游资讯

 吃

海鲜

雁荡山特色佳肴以海鲜为主，当地著名的八大名菜为鸡末香鱼、蛤蜊豆腐汤（在当地有"天下第一鲜"的美称）、蟠龙戏珠（以雁荡山的溪鳗为主料做成）、美丽黄鱼、碧绿虾仁、清蒸海蟹、雁荡石蛙（夏、秋两季难得的佳肴）、土豆野味煲（野兔与土豆焖制）。

雁荡山地区海鲜品种丰富，乐清湾有贝壳40余种，甲壳类60余种，如缢蛏、泥蛤、蝤蛑（又称青蟹）等，水产品年产30余万担（1担=50千克）。这些水产品成为游山观海、品海鲜的旅游美食。

Tourist Information

Food
Seafood

The specialties of Yandangshan are mainly seafood and the eight famous local dishes including Chicken Mince and Aromatic Fish, Clam and Tofu Soup (locally enjoys the reputation of "the freshest in the world"), Two Dragons Playing with a Pearl (Yandangshan marbled eel-based), Yellow Croakers, Verdure Shelled Shrimps, Steamed Sea Crabs, Yandang Stone Frog (rare delicious food in summer and autumn), and Potato and Game Stew (hare and potato stew).

Yandangshan is rich in seafood. There are more than 40 species of shellfish and 60 species of crustaceans in Yueqing Bay, such as razor clam, geodic and swimming crab (also known as blue crab). The annual output of aquatic products is over 300,000 piculs (1picul=30kg). These aquatic products turn into delicious dishes ready for tourists to enjoy after touring and hiking in the mountains and nearby coastal scenery.

鸡末香鱼
Chicken Mince and Aromatic Fish

蛤蜊豆腐汤
Clam and Tofu Soup

清蒸海蟹
Steamed Sea Crabs

蟠龙戏珠
Two Dragons Playing with a Pearl

美丽黄鱼
Yellow Croakers

小吃

来到雁荡山世界地质公园,不得不品尝一下当地独特的传统风味小吃。许多游客畅游雁荡后都说:"观山景、尝小吃、品海鲜,其乐融融。"雁荡风味小吃讲究色香味,只要观其色、闻其香,顿时胃口大开。著名的有香螺、番薯粉丝汤、雁荡烙饼、米粉丝面、茴香五味豆腐干、绿豆面等,清明时节更可品尝到当地美味的雁荡清明粿。

Snacks

When you come to Yandangshan UNESCO Global Geopark, you must taste the local traditional snacks. After a great trip in Yandangshan, many tourists said "seeing the mountain, tasting the snacks and seafoods, what an enjoyable experience!" The famous snacks include Neptunea Cumingi Crosse, Sweet Potato noodle Soup, Yangdang Pancakes, Rice Noodles, Fennel Five-Flavoured Dried Bean Curd, and Mung Bean Noodles. You can also enjoy the local mouth-watering Yandang Round Green Cakes (Qingmingguo) during the Qingming Festival.

香螺

香螺用清水煮熟即可食用。吃法与田螺不同,不能用嘴去吮,只能用一根牙签,揭开其上紫色的薄片盖,然后用牙签挑出螺肉与膏,再放在醋酱味俱全的调料里蘸用即可。

Neptunea Cumingi Crosse

Neptunea Cumingi Crosse is eatable after dipping into boiling water. The way of eating neptunea cumingi crosse is to use a toothpick to open its purple thin cover and pick up the meat and cream inside. This is followed by dipping them in vinegar and soy sauces to bring out its best taste.

茴香五味豆腐干

茴香五味豆腐干采用精制小方块豆腐干、新鲜猪脚、桂皮、茴香、蒜根、精盐同锅煮,煮的时间越长其味越好,直煮得豆腐干内成蜂窝状。食时加少许辣酱,味道更佳。轻咬一口,香气四溢,入口鲜美。

Fennel Five-Flavored Dried Bean Curd

Fennel Five-Flavoured Dried Bean Curd is made by boiling small pieces of refined dried bean curd, together with fresh pig feet, cinnamon, fennel, garlic root, and refined salt. The longer it is boiled, the better it will taste. It is recommended to cook until the dried bean curd becomes honeycomb shaped inside. It tastes remarkable with a little chilli sauce. A small bite will bring out the aroma and freshness.

绿豆面

雁荡山特产绿豆面是用绿豆粉、红薯粉掺和制成的绿色的、筷子粗细的面食,最适合炒着吃。用芹菜、牡蛎、蛏子肉、蛋皮丝炒面,细而不碎、松而不结,十分美味。

Mung Bean Noodles

Another Yandangshan's special dish is Mung Bean Noodles which is made of mung bean powder blended with sweet potato powder. It is a kind of green chopstick like thick pasta that is suitable for frying. The noodles fried with celeries, oysters, razor clam meat, and scrambled egg. Their texture is fine and tough and will not easily break.

番薯粉丝汤

先将番薯加工成番薯粉丝，再把晒干的番薯粉丝与猪脚或肉骨头同煮。食时碗中加入蛋丝、芹菜、炊虾、紫菜、蛎肉、米醋、酱油、味精等配料，吃起来软滑爽口。喜辣的加入少许辣酱，则另有一番滋味。雁荡人宴请宾客、婚嫁丧聚摆酒席都少不了番薯粉丝汤这道菜。

Sweet Potato Noodle Soup

Sweet potatoes are used to make noodles. The noodles are then cooked with pork feet and bones. When ready to eat, add egg, celeries, shrimps, seaweeds, oyster meat, rice vinegar, soy sauce, monosodium glutamate and other ingredients. It tastes soft and refreshing. Chili can be added to make it spicy. Sweet potato noodle soup is an indispensable dish for the banquets, weddings and funerals of Yandangshan people.

雁荡烙饼

雁荡山的大街小巷都摆放着这样一种小摊子：一个火炉，一个平底锅，它们是用来制作烙饼的。雁荡烙饼最讲究的是调料，一般是用上等的咸菜，和着豆腐干、虾皮、肥猪肉、蛋皮一起揉进面粉团里，然后用手捏成薄薄的、圆圆的一片，放在平底锅上烤，三五分钟过后，一张色香味俱全的烙饼便告成功了。

Yangdang Pancakes

The pancake street stalls in Yandansghan all have their own stoves, flat bottom pans to make pancakes. Yandang Pancakes use ingredients such as finest pickles, dried bean curd, shrimps, pork, and egg kneaded into dough. The dough is then rolled to form thin and round shape with hands. It is put into the pan and baked for three or five minutes to a pancake with great colour, aroma and taste.

米粉丝面

雁荡小吃中还有一种当地人称之为"细面"的米粉丝，同样是用大米做的粉丝，它却细如松针，细而不碎，黄而不焦，松而不结，滑而不腻，加上牡蛎、蛏子肉、蛋皮丝，的确会令人嘴馋。

Rice Noodles

The local "fine noodles" are another Yandang snack, which are made of rice. They are as thin as pine needles but not easy to break. They carry a yellow colour but not burnt, loose but not knotted, smooth but not greasy. With oysters, razor meat, and egg slices, the noodles are one of the most favoured mouth-watering snacks.

住 / Hotel

雁荡山世界地质公园各园区均有星级酒店、宾馆、民宿供您选择。

There are star-rating hotels, guest houses and homestays for you to choose from in Yandangshan UNESCO Global Geopark.

宾馆名 Hotel Name	联系电话 Telephone No.	地址 Address
雁荡山山庄 Yandangshan Villa	0577-62245333	雁荡山霄霞路8号 No. 8, Xiaoxia Road, Yandangshan
银鹰山庄 Yinying Villa	0577-62246666	雁荡山霄霞路7号 No. 7, Xiaoxia Road, Yandangshan
假日大酒店 Holiday Hotel	0577-62245555	雁荡山宵霞路6号 No. 6, Xiaoxia Road, Yandangshan
曙光山庄 Shuguang Villa	0577-62245222	雁荡山响岭头村步行街151号 No. 151, Walking Street, Xianglingtou Village, Yandangshan
温岭九龙国际大酒店 Wenling Jiulong International Hotel	0576-81688888	温岭市万昌中路688号 No. 688, Wanchang Middle Road, Wenling City
方山云顶 Fangshan Yunding Hotel	0576-81609801	方山景区内 In the Fangshan Scenic Area
永嘉县宾馆 Yongjia Hotel	0577-67220888	永嘉县上塘镇县前路85号 No. 85, Xianqian Road, Shangtang Town, Yongjia County
永嘉林溪小院客栈 Yongjia Linxi Small Courtyard Inn	0577-57668388	永嘉县枫林镇兆潭村 Zhaotan Village, Fenglin Town, Yongjia County

旅游服务
Tourist Services

交通状况

　　雁荡山世界地质公园交通便利，距杭州300千米，距温州70千米，104国道、沈海高速公路及甬台温高速铁路直达公园，北有台州路桥机场，南有温州龙湾国际机场和温州港，形成进入公园的海、陆、空交通网络。

Transportation

Traffic Conditions

　　The access to Yandangshan UNESCO Global Geopark is convenient. It is 300km from Hangzhou and 70km from Wenzhou, connected by the China National Highway 104, Shenyang-Haikou Expressway and Ningbo-Taizhou-Wenzhou Railway. It is also close to Taizhou Luqiao Airport in the north, and Wenzhou Longwan International Airport and Port of Wenzhou in the south. The geopark is well-connected to a comprehensive sea-land-air transportation network.

雁荡山世界地质公园交通位置图
Transportation Network of Yandangshan UNESCO Global Geopark

火车

雁荡山火车站距离雁荡山世界地质公园仅6千米，各主要城市均有高铁经过雁荡山，方便快捷。

Train

The Yandangshan Station is only 6km away from the Yandangshan UNESCO Global Geopark. There are high-speed trains link up Yandangshan with other major cities.

| 温州南站 Wenzhou South Railway Station | | 30分钟 雁荡山站 Yandangshan Railway Station |

| 杭州站 Hangzhou Railway Station | | 2小时30分钟 雁荡山站 Yandangshan Railway Station |

| 上海虹桥站 Shanghai Hongqiao Railway Station | | 3小时30分钟 雁荡山站 Yandangshan Railway Station |

| 南昌西站 Nanchang West Railway Station | | 6小时24分钟 雁荡山站 Yandangshan Railway Station |

| 宁波站 Ningbo Railway Station | | 1小时40分钟 雁荡山站 Yandangshan Railway Station |

| 福州南站 Fuzhou South Railway Station | | 2小时30分钟 雁荡山站 Yandangshan Railway Station |

| 南京南站 Nanjing South Railway Station | | 4小时 雁荡山站 Yandangshan Railway Station |

| 合肥站 Hefei Railway Station | | 6小时50分钟 雁荡山站 Yandangshan Railway Station |

汽车

雁荡山汽车站距离雁荡山世界地质公园仅3千米，104国道、甬台温高速公路途径雁荡山。

Bus

The Yandangshan Bus Station is only 3km away from the Yandangshan UNESCO Global Geopark, and State Road 104 and Yongtaiwen Expressway also pass through Yandangshan.

| 温州市新城客运中心 Wenzhou Xincheng Passenger Transport Centre | | 1小时30分钟 雁荡山客运中心 Yandangshan Passenger Transport Centre | 流水班 Non-fixed schedule |

旅游服务
Tourist Services

飞机
您可以选择飞往台州路桥机场或温州龙湾国际机场再乘坐火车或汽车直达雁荡山。

Airplane
To get to Yandangshan, you can fly to Taizhou Luqiao Airport or Wenzhou Longwan International Airport and then take the direct train or car.

出发地	到达地	时长
北京 Beijing	温州 Wenzhou	2小时40分钟
北京 Beijing	台州 Taizhou	2小时50分钟
上海 Shanghai	温州 Wenzhou	1小时25分钟
广州 Guangzhou	温州 Wenzhou	1小时50分钟
广州 Guangzhou	台州 Taizhou	2小时50分钟
贵阳 Guiyang	温州 Wenzhou	1小时55分钟
西安 Xi'an	温州 Wenzhou	2小时35分钟
成都 Chengdu	温州 Wenzhou	2小时30分钟

游
来到雁荡山如果您不是自由行，那么您可以报团旅行。公园周边有丰富的旅行社供您选择。

Travel
If you do not prefer to travel alone in the geopark, you can join a tour. The following are travel agencies.

旅行社名称 Travel Agency Name	联系电话 Telephone No.
雁荡山旅游总公司 Yandangshan Tourism Head Office	0577-62241488
雁荡山朝阳旅行社 Yandangshan Chaoyang Travel Agency	0577-62241938
雁荡山山河旅行社 Yandangshan Shanhe Travel Agency	0577-62243244
雁荡山环宇旅行社 Yandangshan Huanyu Travel Agency	0577-62243436
温岭市旅行社 Wenling Travel Agency	0576-86595717
浙江假日国际旅行社 Zhejiang Holiday International Travel Agency	0576-81623338

购

以下特产均可在当地旅游商店买到。

Shopping

All the above local produces can be purchased at the local tourist shops.

海鲜干货

公园所在地区的海产品丰富,当地人喜欢将鱼虾、贝类制作成干货,作为馈赠亲友的佳品。

Dried Seafoods

Abundant seafoods are available locally. Fishes, prawns and shellfish are dried for the ease of bringing back as souvenirs by tourists.

香鱼

细鳞、味美无腥,产仔于潮滩,鱼苗由海洄游到涧溪。

Aromatic Fish

With fine scales, delicious and not fishy, aromatic fish are abundantly found in the tidal flat, with fish seedlings migrating from the sea to rivers and streams.

雁茗

雁山茶,称雁荡毛峰、雁荡云雾,历代对雁山茶赞赏颇多。 朱谏《雁山志》载:浙江多茶品,而雁山茶则称最。清代端木国珊著有42句长的《雁荡山茶歌》,宋、明、清三代均将雁山茶作为贡品。

Yan Tea

Yan Tea is the tea grown in Yandangshan. It is also called Yandang Maofeng or Yandang Cloud Tea. The tea has been famous and complimented for dynasties. As recorded in *Yanshan Annals* written by Zhu Jian, there were various types of tea in Yandangshan but the best is Yanshan Tea. Duanmu Guoshan in Qing Dynasty wrote the *Song of Yandangshan Tea* with 42 sentences. Yanshan Tea was taken as a tribute tea for the Emperor in Song, Ming, and Qing Dynasties.

铁皮石斛

铁皮石斛为兰科石斛兰,属多年生草本植物,有益胃生津、滋阴清热的功效,是名贵的中药材、养生极品。有野生兰花生长的地方才有可能良好的生长,而雁荡山恰好是个产兰花的地方。紧邻雁荡镇的大荆镇有很大的铁皮石斛种植生产基地。

Dendrobium Officinale

Dendrobium officinale of Orchidaceae Dendrobium is a perennial herb. It is good for stomach, can generate body fluids and positive energy. It is a valuable Chinese herbal medicine and one of the best health supplements. It only grows well at places where wild orchids grow, and Yandangshan happens to provide a perfect habitat to grow orchids. There are very large *Dendrobium officinale* cultivation farms in Dajing Town close to Yandang Town.

旅游服务
Tourist Services

常用电话
Frequently Used Telephone Numbers

游客服务中心 Tourist Service Centre	0577-62242003
旅游服务专线 Travel Service Hotline	0577-62178888
门票结算中心 Ticket Centre	0577-62242071
投诉电话 Complaint Hotline	0577-62243651
报警中心 Police Centre	110、0577-62243139
火灾报警 Fire Report Centre	119、0577-62243656
医疗急救 First Aid	120
交通事故 Traffic Accident	122
消费投诉 Consumer complaint	96315

旅游小贴士
Travelling Tips

1. 雁荡山世界地质公园最佳旅游季节为春、夏、秋三季。
   The best seasons for touring in Yandangshan UNESCO Global Geopark are spring, summer, and autumn.

2. 公园雁荡山园区灵岩景区每天上午和下午安排有飞渡表演，请游客注意安排时间。
   Please schedule your time in order not to miss the Fly Across Show conducted every morning and afternoon in the Lingyan Scenic Area of Yandangshan Scenic District.

3. 游览雁荡山，一定要去灵峰景区看夜景，这里是雁荡山变幻多姿的精髓所在。
   When visiting Yandangshan, the night scene of Lingfeng Scenic Area is a must to see.

4. 游览楠溪江园区，竹筏漂流是您不可错过的项目之一，在渡头、二桥、狮子岩、龙河渡均有不同河段的漂流服务供您挑选。
   When visiting Nanxijiang Scenic District, you can not miss the bamboo rafting. Services are available in Dutou, Erqiao, Lion Rock, and Longhe Piers.

5. 在漂流的过程中，务必不要在竹筏上嬉戏打闹，防止落入水中。
   To avoid falling into water, don't play when travelling on bamboo raft.

6. 无论是什么季节，一定要配备好防晒用品及一些简单的药品，以备不时之需。
   It is recommended to bring sunblock and personal medicine just in case in any seasons.

7. 您可以乘坐雁荡山旅游公交到达雁荡山园区的各个景区。要注意，最好在17:00前回到公交站乘车返回。
   You can visit different scenic spots by shuttle buses in Yandangshan Scenic Area but please ensure you arrive at the bus stops before 5 p.m.

8. 游览长屿硐天时，请合理安排时间，别错过每天在观夕硐内的岩硐音乐表演。
   Please arrange your time when visiting Changyu Dongtian and do not miss the Cave Concert performed every day in Guanxi Cave.

9. 方山景区有宾馆及民宿供您选择，可以游览，等待早晨壮美的日出盛景。
   There are hotels and homestays in Fangshan Scenic Area for you to choose from. You can enjoy your tour at ease, waiting for the magnificent sunrise in the morning.